HIGH DATA RATE TRANSMITTER CIRCUITS

THE KLUWER INTERNATIONAL SERIES IN ENGINEERING AND COMPUTER SCIENCE

ANALOG CIRCUITS AND SIGNAL PROCESSING
Consulting Editor: Mohammed Ismail. *Ohio State University*

HIGH DATA RATE TRANSMITTER CIRCUITS

RF CMOS Design and Techniques for Design Automation

by

Carl De Ranter

RF MAGIC, U.S.A.

and

Michiel Steyaert

KU Leuven, Belgium

KLUWER ACADEMIC PUBLISHERS
BOSTON / DORDRECHT / LONDON

A C.I.P. Catalogue record for this book is available from the Library of Congress.

ISBN 978-1-4419-5381-0   e-ISBN 978-0-306-48713-2

Published by Kluwer Academic Publishers,
P.O. Box 17, 3300 AA Dordrecht, The Netherlands.

Sold and distributed in North, Central and South America
by Kluwer Academic Publishers,
101 Philip Drive, Norwell, MA 02061, U.S.A.

In all other countries, sold and distributed
by Kluwer Academic Publishers,
P.O. Box 322, 3300 AH Dordrecht, The Netherlands.

Printed on acid-free paper

to Ans,
in remembrance of
Erik Grobben

Contents

List of Figures

List of Tables

Chapter 1

INTRODUCTION

> "In recent years an exponential growth could be seen in the market of personal communications, ..."

Although this statement was true in 2000, only two years ago, now it can be considered as a part of history. The cyclic behavior of the market of consumer electronics is often forgotten when the period of seven good years has arrived. It is mostly experienced as a period of "unpreceded growth" and "unimaginable possibilities". However, the economic law for a cyclic industry inherently holds a period of reduced or even negative growth. Although, or maybe due to the fact that it is clear today a period of downturn has arrived for the electronics industry, the pace at which system demands for electronic devices are getting tougher, is still increasing whereas the price one can ask for such systems is steadily dropping.

Especially for analog RF design, the challenge to design fully functional systems at an economically profitable scale will become bigger and bigger. The design team that tries to beat the odds, thus realizing these electronic systems with higher performance for a lower total cost will surely succeed in exposing itself to an enormous economic pressure and a high work load. Success, however, is not guaranteed at all. Possibly even more important than hard work to reach success are decisions made on a higher level. These encompass choices in design methodology, in chip technology, in manufacturing technology, in software environment etc. Obviously, it goes beyond the scope of this text to thoroughly discuss all of these topics.

This work will focus on the design aspects of electronic circuits that can be used in systems for high-speed data transmission. Some aspects about transistor and inductor modeling and the manufacturing technique of flip-chip bonding will be discussed. It will also be shown in this work how software tools can help shorten the design time of RF building blocks and of CMOS fully-integrated voltage-controlled oscillators in particular.

In a first section of this introduction, some observations about digital and analog chip design are discussed. A second section handles about the important choice of technology that is at hand at the start of every new project involving chip design. Then, an overview is given of the topics dealt with in this work.

1.1 Some Observations

To start this report on a few years of research, a basic ingredient of the scientific approach is used: the observable fact or observation. Here are five of them:

Observation1:

> The fabrication facilities (in short called *fabs*) for the newest CMOS technologies suffer from exponentially rising installation costs resulting in a very high cost per chip area in the startup phase of a new technology.

This is clearly illustrated in Table 1.1 by the cost per mm^2 for the MPW (Multi Project Wafer) service provided by Europractice[1] (www.imec.be/europractice). Moreover, the area cost of a mature CMOS technology also increases for decreasing minimal feature size due to the high manufacturing cost of the mask sets.

Table 1.1.   Silicon cost per mm^2 for MPW-runs of Europractice

Technology	Cost 2001 Euro/mm^2	Cost 2002 Euro/mm^2
UMC 0.25μm RF-CMOS	1700	454
UMC 0.18μm RF-CMOS	2600	1000
UMC 0.13μm RF-CMOS	16500	3980

Thus, the total cost of a chip designed in the latest CMOS technology may suffer from the large area consumption of passive components as e.g. on-chip inductors, because this can lead to a global system cost being higher than the total cost of a system with external inductors. Therefore, the use of CMOS for the design of technologically advanced systems for newly-to-develop markets might not be the best solution what global cost management is concerned.

---

[1]The IC service of Europractice offers low-cost ASIC prototyping and ASIC small volume production through Multi Project Chip - MPC - and dedicated wafer runs

Observation2:

First-time-right silicon is becoming one of the most stringent demands to come to a successful product development.

If a full mask-set redesign is needed, the *Non-Recurring Engineering (NRE)* cost increases significantly. Moreover, the delayed market introduction due to the time needed for this redesign leads to a lower introduction price or the loss of major customers. The combination of both negative effects of a redesign, often suffice to degrade that high profit-promising new product to one introducing loss instead.

Observation3:

Digital designs still benefit from the shrinking gate lengths of new CMOS technologies.

The decreasing supply voltage results in a decreasing power consumption, and the decreasing device sizes result in a lower area consumption. However, very-large scale integration introduces new problems that can only be solved at the penalty of a lower area density. E.g. on large chips, new clock distribution schemes result in a larger area overhead for the clock distribution. Also, the decreasing threshold voltages result in increasing leakage power that has to be dealt with on circuit level, again resulting in a certain area overhead. For analog design, the decreasing supply voltage reduces the available voltage swing, resulting in general in worse signal-to-noise and signal-to-distortion ratios. Moreover, the shortening channel lengths pose new problems for the transistor modeling.

Observation4:

The increase in digital design productivity has been made possible by introducing an abstract representation of the physical signals at the digital gates.

By representing these signals on a higher level as strictly "high" (1) or strictly "low" (0), the design on the levels above becomes a problem that can be solved on a strict mathematical way, using boolean logic. It has lead to a fully-automated top-down design approach using CAD tools that are based on a boolean equivalence between two hierarchical levels.

In analog circuitry, this abstract representation of physical signals can only be made if the error due to the abstraction is also taken into account. For high-performance designs, this often means that the complexity of the abstract representation is not lower than the original complexity. Tools that try to implement the fully-automated design approach for analog circuits will certainly have to take errors into consideration starting from the lowest design level, as demonstrated in [Daems TCASI99].

Observation5:

> Most CAD tools for analog design that are being published try to implement a fully-automated design environment.

This means that the number of tools that support a designer in his pursuit for new, high-performance topologies are not so numerous. However, those few existing tools do improve design productivity in a quite convincing manner [vdPlas PhD01]. Therefore, it is to be expected that the introduction of interactive design-aids that can be used by the designer during his topology exploration and that include a fluid transition to an automated design optimization will push the overall productivity of analog designers towards a higher level.

1.2   To CMOS or not to CMOS

For electronic devices aimed at the consumer market which is characterized by high volumes but low profit margins in a mature market, that is the question indeed. Up till now, CMOS is regarded as the technology that eventually leads to the lowest (marginal) cost per unit in production. Since the rising area cost of the CMOS technologies can only result in a higher profit margin on the final product if the functionality (and the perceived value[2]) of the design increases accordingly, the question has to be answered separately for digital and analog:

- For digital, the functionality generally increases with the number of transistors on a single die. Also, no real limit exists for the total number of transistors on one chip since the different functional cores present on large systems-on-chip do not interfere with each other. Here, the problems of heat production (cooling), clock distribution and leakage power reduction are the challenges the digital designer will have to deal with;

- For analog, the picture is completely different. Here, the decreasing supply voltage and the lower intrinsic linearity ([Borr JSSC98]) or the lower power efficiency for a certain linearity of MOST devices ([Jans PhD01]) with shrinking channel lengths might limit a further downscaling of an analog circuit. Power efficiency is important to obtain a low power consumption for battery-operated circuits or to minimize the heat production of the chip ([Pies ISSCC01]).

Therefore, it can be said that for digital it is mostly advantageous to transfer a design to the newest fully-characterized technology. For analog this will not always be the case because of the problems mentioned above, and also because of the increasing cost per area of non-scaling elements as on-chip inductors and

---

[2]This is in fact a good indication of the price a consumer might want to spend to buy the product.

decoupling capacitors. These non-scaling elements could even render the total chip cost higher in the "smaller" new technology than in the old one.

Recently, it has been uttered that ([Wils EE02]):

> "[...] the SoC-based system simply becomes a delivery vehicle for revenue-producing services."(J. Tully, Dataquest)

These SoC (System-on-Chip) systems will in fact provide wireless connectivity with high data rates using a plethora of different protocols. Due to the increasing level of integration, more and more problems on the RF system level will have to be dealt with on the integrated system itself instead of on the board the chip is implemented on. This will necessitate a clear understanding of all interface-related problems between on-chip and off-chip signals, and a close cooperation between the RF system engineers and the chip-design engineers. Among these interface-related problems the availability of good package models certainly has to be included. Good device models are needed to avoid expensive redesigns and to shorten the time to market. To additionally shorten the time-to-market of these integrated systems, the analog designer will have to count on tools that speed up "simple" and repetitive design tasks. To further increase analog design productivity, tools can be envisioned that enable the optimization of tightly entangled analog building blocks based on the know-how of the designer, thus decreasing the work load of "manual", try-and-error-based design optimization.

Returning to the quote above and looking back at the previous boom of hand-held devices for personal communications, one can say that the introduction of these new RF systems will add a third player to the "old" duo of technology provider *(building the devices)* and voice service provider *(building the network)*. This third player will be the provider of the services that are being delivered over the network and that are running on the device of the end-user. According to the quote, this third party can probably count on the highest profit.

However, someone still has to make and sell those millions of devices. And here the importance of the right technology choice comes into play. Probably, a mature CMOS technology is still the cheapest if only the marginal cost[3] is taken into account. However, CMOS still suffers from longer development times than e.g. bipolar designs. Also, CMOS transistor models are still less reliable for RF simulations than their bipolar counterpart. These two drawbacks can delay the market introduction of the final product. So, although the marginal cost of the product has been minimized by using the cheaper CMOS technology, the problem could pose itself that both profit margin and market share are low once the production is finally ramped up to high volume. It does not need further explanation that this is where the story ends.... Obviously, the use of design tools to shorten the CMOS design cycle could help to avoid this unfortunate ending.

---

[3]The effective cost per additionally manufactured chip.

1.3   Covered Topics

The technical realizations presented in this work are situated in the area of high-speed data communication circuits. The CMOS technologies that have been selected for these designs still result in the lowest marginal cost. However, due to the longer design time and higher risk as compared to a similar design in a bipolar technology, they do not provide the maximum overall revenue for an industrial design project, thus justifying the need for this work to be done in a research environment.

As mentioned in the previous section, a lot of issues pop up during the design of an integrated circuit. In a first chapter, the influence of substrate resistivity is discussed. This will help to choose the right "flavor" of a certain CMOS technology to integrate RF circuits. Reliable device models are needed to avoid expensive redesigns. Therefore, a description of the device-modeling approach as used for the chip designs presented in this work is also included in the first chapter. For both MOST devices and on-chip inductors a practical model that is still accurate enough to prevent the need for redesign, is derived from literature and a detailed study of the device layout. Also, the need for accurate quadrature signals in transmitters is explained and two approaches to generate such a quadrature signal are described. Finally, the first chapter is concluded by a discussion of some manufacturing issues with an emphasis on the flip-chip bonding technique that has been used for the upconvertor presented in chapter five.

To reduce the design time of (CMOS) analog circuits, design tools are needed that help the designer in his job. In chapter three a possible framework for analog CAD is presented that is based on a topology-specific Design Template with matching Design Directives. This framework might be the basis for an environment that supplies an analog designer with the productivity-enhancing tools he needs. As an example of the use of this kind of framework for the design of high-performance RF designs, the CYCLONE tool for automated design and layout of voltage-controlled oscillators (VCO) is described. It has been used to design a 3.3 GHz VCO in a $0.35\mu m$ CMOS technology with a low substrate resistivity.

Chapters four and five describe in detail the circuits and measurement results of chip realizations in different CMOS technologies. In chapter four, the design of voltage-controlled oscillators for both broadband systems and high-frequency applications is discussed. It will become clear how differences in system specifications lead to a completely different focus on the requirements of even a very basic building block as an oscillator. For the broadband system, a four-stage ringoscillator has been integrated in a $0.5\mu m$ CMOS technology. Operating from 3 V the oscillation frequency can be tuned between 55 MHz and 1.2 GHz, being more than a decade. Using a polyphase network, the quadrature oscillator output is linearized in order to reduce the harmonics that would fall within the band of interest. In a $0.25\mu m$ CMOS technology, a high-frequency LC-tank type VCO has been realized that oscillates at 17 GHz, while still having a tuning range of 8.6%. The measured phase noise is -108 dBc at an offset of 1 MHz from the carrier frequency.

In a fifth chapter, the design of an upconvertor for high-bandwidth data signals is presented. The chip has been manufactured in a $0.25\mu m$ CMOS technology. It can transmit baseband signals with a bandwidth of over 15 MHz at carrier frequencies between 900 MHz and 2 GHz with a linearity that is sufficient for higher-order digital modulation schemes. Operating from a 2 V power supply, the transmitter consumes 25 mW to provide an output power of -12 dBm at 2 GHz and -10 dBm at 900 MHz. This chapter can be read as a typical design case, since it starts from a brief market perspective, followed by a high-level topology design and subsequently a (qualitative) derivation of the block-level specifications. After a description of the final circuits and the used design methodology, the measurement results are presented.

Chapter six concludes this work.

Chapter 2

RF MODELING, QUADRATURE GENERATION AND FLIP-CHIP BONDING

2.1   Introduction

In modern CMOS technologies often the designer is offered the choice between a substrate with high or one with low resistivity. In a first section the advantages of substrates with high resistivity are explained. It is shown that not only on-chip inductors within an analog design benefit from this high-resistivity substrate. Also mixed-signal designs as digital-to-analog convertors and fully digital circuitry can obtain better performance when a technology with high substrate resistivity is chosen.

A successful design of analog RF circuitry strongly depends on the reliability of the models for both active and passive components. Therefore, in a second section a short overview is given of the models that have been used in the designs presented in Chapter 4 and Chapter 5 for the RF MOSFET transistors and for the on-chip inductors. The need for a good quadrature local oscillator signal for upconversion circuits is explained in a third section. The two mostly used approaches to obtain such a quadrature oscillator signal are also discussed in that section.

Packaging is another important issue when dealing with RF designs. As with other parasitics, a good strategy is to incorporate from the beginning also the parasitics coming from the final chip package into the design. In deep submicron design, these parasitics might become predominant compared to the on-chip parasitics. For instance, a bondwire connection inserts a series inductance of 1 to 5 nH, resulting in an impedance up to 75 Ω at 2.4 GHz. A solution to avoid this bondwire and its inherent inductance, is to use flip-chip bonding. This technique, with its advantages and possible assembly problems is discussed in the fourth section. In the conclusions, a brief overview is given of the way the elements discussed in this chapter are used in the remainder of the text.

2.2   Influence of Substrate Resistivity on RF-designs

The choice of the right technology for a specific project is becoming more and more a distinctive factor in the global profitability of the final product. It is not obvious at all that plain CMOS is the right technology-choice for high-performance RF circuits, but in the case it is chosen after all, often a choice of substrate resistivity is offered by the chip foundry. In this section, the advantages of a high-resistivity substrate for the performance of both analog and digital circuits is explained. An LC-tank VCO is a typical RF circuit that benefits from a high substrate resistivity. This is demonstrated in a short design case wherein two VCO's are compared.

2.2.1   A statement to start with...

Let's kick off with a very simple statement that expresses quite well how the use of an epi-layer can be evaluated:

"EPI=EPO for digital"

This, of course, needs some explanation. Let's take a look at the abbreviations involved. First of all, there's EPI. This stands for epi-*layer*, and any reader acquainted with the topics discussed in this book will probably also know that an epi-layer is the low-conductive top layer grown on the wafer of almost all the CMOS technologies of the previous generation. However, the word "EPO" does need some explanation. For that, we have to go to the medical world. It is a growth factor that stimulates the production of blood cells and it became very popular among professional athletes some years ago.

The result of using EPO was that the performance of these people got a significant boost, as is mostly the (positive) effect of a dopant. A competitor not using the dopant, has no benefit from it and his job becomes more difficult. For electronic design, the introduction of an epi-layer on top of a low-resistivity substrate, boosted the integration density of digital circuits. The layout density of (standard) cells in a digital cell library effectively rises when a low-resistivity substrate is used since less substrate straps are needed to keep the resistance to ground low enough in order to avoid the latch-up effect.

However, for the analog and mixed-signal designs the use of this low-resistivity substrate imposes some problems:

- The coupling between digital and analog building blocks through the substrate is much higher, resulting in higher substrate noise [Arag JSSC99]. E.g. delta-sigma convertors, were said to have a lower performance what their dynamic range is concerned, on low-resistivity substrate when compared to an identical design on a high-resistivity substrate [Mori ESSCIRC99];

- For high-frequency designs, especially those involving on-chip inductors, induced substrate currents result in a power loss and a reduced efficiency of the system ([DeRa TCAD02]). This effect is dealt with in more detail further on.

But, a little in contrast with the reason of existence of the wafers with epi-layer, also for exclusively digital circuits the use of a high-resistivity substrate can be beneficiary. This is explained in the next section.

2.2.2 A High Substrate Resistivity for Digital?!

Nowadays clock frequencies within digital chips are rising to levels at which high-frequency effects like skin effect, eddy currents and substrate coupling are not negligible anymore. Possibly even transmission line behavior has to be taken under consideration for long on-chip connections. Thus, the hard line that could be drawn between analog and digital signals is becoming more and more diffuse. And then the question arises: *"Will the use of low-resistivity substrates in the end limit the performance of digital circuitry ?"* The parallel with a medical dopant is mind-boggling: their use tends to boost the physical performance of the athlete, but at the same time it undermines the health of the person at later age.

The answer to the question acknowledges the parallel with the medical world. First of all, the scaling of technology already enables the integration of such huge digital systems on a single die, that the quest for the most compact digital cell library no longer prevails. Other problems in the layout of digital systems become more of a challenge to deal with. One of them is the influence of the high-frequency effects mentioned above on the overall delay. In [Kol'd SISPAD98] the effect of substrate coupling on the effective resistance of a metal interconnect has been modeled. Fig. 2.1 shows the effective resistance of a straight metal line with a width of $10\mu m$, thickness of $0.5\mu m$ and an oxide thickness between metal and substrate of $3.1\mu m$, plotted against the frequency of a signal sent over this line. The plot is shown for two different substrate resistivities.

The high-resistivity substrate shows a constant metal line resistance for frequencies up to 1 GHz, whereas the low-resistivity substrate already shows a big increase in metal line resistance starting from 100 MHz. The induction of eddy currents in the substrate is the main cause of this increase in resistance. This increase can also be expressed as an increasing delay over the metal line. For digital systems, any extra delay results in a lower maximum clock frequency or an increase in power usage and latency due to the need for extra latches to shorten the critical path. It can be concluded that for digital chips, the use of a substrate with a high resistivity will lead to a better performance at the cost of a slightly decreased area efficiency.

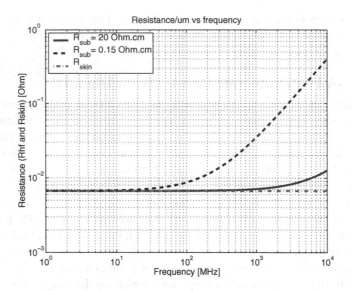

Figure 2.1.   Resistance of a metal line on chip

2.2.3   VCO's and Substrate Resistivity

The previous section dealt with the influence of the substrate resistivity on the performance of high-speed digital chips. For systems-on-chip, consisting of both analog and digital building blocks, the importance of a high substrate resistivity comes from the inherent substrate noise isolation it provides. In this section, it is clarified how *analog* circuit design has a direct advantage on the use of a technology with a high-resistivity substrate, and why this advantage becomes even more important for higher on-chip frequencies. More specifically this is done by looking at the influence of the substrate resistivity on the electrical parameters of an on-chip inductor.

For this, a comparison is made between an analog building block built around such an inductor, simulated in two "flavors" of an existing technology. The first one is an existing 0.25μm technology with a high-resistivity substrate whereas the second one is "virtual" in the sense that it only differs from the first one by its *lower* substrate resistivity. The building block that is chosen for this small case-study is a voltage-controlled oscillator (VCO) with an LC-type topology implementing an on-chip inductor, as shown in Fig. 2.2. This type of VCO is one of the most elementary building blocks that uses an inductor and that can be used to evaluate its quality factor. This quality factor depends on all losses the current flowing in the inductor is subject to. The modeling of these high and low frequency losses is described in the next section (Section 2.3.3.2). The discussion there also shows a proportionality of these losses with frequency.

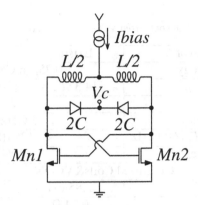

Figure 2.2. LC-VCO topology used for comparison

The CYCLONE tool, described in Chapter 3, has been used here to obtain an optimized VCO design, for both "flavors" of the used technology. The optimization is set up to minimize the power usage for a VCO with the topology of Fig. 2.2, with a specified maximum phase noise and a specified minimum tuning range. This way, the power usage of the optimal VCO in each technology flavor can be used as a measure for the quality of the inductor. In more general terms, the power usage of the VCO can be used as a measure for the suitability of the technology for the implementation of RF building blocks. In the top half of Table 2.1 the design specifications for the two VCO's are given. The bottom of the table depicts the parameters of the resulting optimized VCO's. Obviously, the VCO designed in the technology flavor with low substrate resistivity needs more power to obtain the comparable specifications as the VCO in the technology flavor with a high substrate resistivity. This clearly shows that for analog RF building blocks, a better performance is to be expected when using a technology with a high substrate resistivity, as compared to one having a low substrate resistivity. These conclusions are completely consistent with the measured effects of the substrate resistivity on the quality factor of on-chip planar inductors as reported in [Long JSSC97].

2.2.4 Conclusion

From the discussion held in this section, it should be evident for the reader that in conclusion a pronounced preference for substrates with a high substrate resistivity can be formulated. This is especially true for high performance designs at RF frequencies. The common belief that a high substrate resistivity is beneficiary for analog designs, is supported once more by the demonstration of a simulation example. However, another common belief that *only* analog blocks benefit from the use of a high-resistivity substrate is countered. It is clarified that also mixed-mode designs and exclusively digital designs will need high-resistivity substrates to keep improving the global system performance.

Table 2.1.   Technology used and optimized VCO's for each flavor

Tech. flavor	Existing	"Virtual"
Technology	0.25 μm CMOS	
$\rho_{Substrate}$	14.3 Ω.cm	0.0143 Ω.cm
Vdd	2 V	
Frequency	2 GHz	
Phase Noise @ F_c+600 kHz	< -120 dBc	
Tuning Range	> 20%	
Optimized Coils&VCO's		
Ls	2.85 nH	1.81 nH
Rs	0.74 Ω	0.95 Ω
Power	8.8 mW	12.8 mW
Phase Noise (calc.)	-120.05 dBc/Hz	-120.9 dBc/Hz

2.3   RF-Modeling

2.3.1   Introduction

Already for some years, the design of the analog partition of systems-on-chip suffers from a much lower efficiency in terms of needed design resources relative to the final chip area as compared with the digital partition, for which automated design and layout tools increased the design productivity with some order of magnitude. As is commonly known, this increase in design productivity has been enabled by making abstraction of the physical signal levels at the digital gates, representing these signals on a higher level as strictly "high" (1) or strictly "low" (0). The physical behavior of a digital cell block is reduced to the introduction of a certain time delay on top of its logical function. This kind of abstraction at gate or cell block level is impossible in the design of analog building blocks. Therefore, the design of the analog partition is still based on circuit simulations incorporating physical models of the used devices to predict the signal behavior of the manufactured chip.

Due to the rising production costs of silicon chips and the ever shortening product cycle time, the demand for first-time-right silicon is becoming one of the most stringent demands to come to a successful product development. The combination of high development costs due to the need for a redesign and a lower introduction price due to a delayed market introduction time, often suffice to degrade that high profit-promising new product to one introducing loss instead of profit. Therefore it is clear that also the analog partition has to fulfill all specifications without the need for a redesign. Since the analog performance is directly dependent on the physical signal behavior in the chip after manufacturing, an accurate prediction of this signal behavior in the design stage of the chip is extremely important. This is only possible if good device models for both active and passive components are available at design time. It can be concluded that the quality of the device models that are available to the analog designer has an increasing impact on the final profitability of a complete project.

As final introductory remark, one also has to envisage the effects of the down-scaling of CMOS technologies for the digital design. More and more, the abstraction that is made of the physical behavior of digital cell blocks becomes problematic:

- The decreasing threshold levels and power supply voltages lead to problems with available noise margin and can consequently result in yield problems or non-optimal performance when the final level of digital design simulation would remain at the digital cell block level. In some cases, a more optimal digital design might be obtained by descending into the cell and performing transistor-level simulations as is typically done in analog design;

- The decreasing threshold levels also lead to increasing leakage power which has to be managed on circuit level;

- The increasing clock frequencies combined with larger chips due to the integration of more complex systems lead to electrical line lengths on chip that no longer permit to neglect all transmission line effects during the design.

The trend at this moment in digital design is not to incorporate these effects in the final system simulations, but instead to divide the system in subsystems for which the conditions that allow for abstraction of the physical behavior are still valid. However, for the partitioning of those subsystems and for the exact modeling of the digital cell blocks, the physical reality does have to be taken into account and therefore also in digital design proper RF device modeling will become more and more important.

In this section, first the modeling of MOST devices is dealt with. In a second subsection, a simulation strategy for on-chip inductors, based on an automatically generated lumped model, is explained.

2.3.2 MOS Transistors

2.3.2.1 Introduction

For the simulation of the behavior of a MOSFET device different compact models like BSIM3 ([BSIM3]), EKV ([EKV]) and MOS Model 9 ([Model9]) exist. Of these models, the BSIM3v3 model can be considered as the *de-facto* standard because it is the model that is generally provided by the silicon foundries. For most low-frequency applications [1], all compact models are quite appropriate to predict the final behavior of a circuit after manufacturing. However, the compact models fall short in the modeling of high-frequency effects. For the BSIM3 model, another problem is that not all parameters of the model are provided in the technology files as distributed by the silicon foundries. This causes the MOSFET models a designer has to work with to lack some elements that may have an important influence on the transistor behavior at high frequencies. In this context, "high frequencies" means in the order of 1/10th of the transistor's cut-off frequency f_T. For a standard 0.25μm CMOS technology with an f_T of around 20 GHz for practical values of $V_{GS} - V_T$, a few gigahertz thus is considered here to be a high frequency.

A first subsection discusses the construction of a subcircuit used in this work to model the high-frequency behavior of a MOS device. It is based on data coming from the technology provider combined with a study of the MOS transistor itself. In a second subsection this model is validated by comparison with a measurement-based model provided by the foundry. A third subsection gives some remarks on the non-quasi-static effect. In a last subsection some frequency-performance indicators that are used to quantify the frequency behavior of a MOS device are presented.

2.3.2.2   Model Construction

Fig. 2.3 shows a cross-section along the channel of a MOS device, with an indication of the different parasitic elements that have to be taken into account. As stated above, not all these parasitics are included in the BSIM3 model parameters provided with a standard design kit.

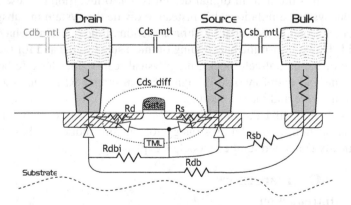

Figure 2.3.   Cross section of a MOSFET device

---

[1]For CMOS in the year 2002 this means below 100 MHz

Following elements are among those that are not or only partially included:

1 The *non-quasi-static effect* is not included in the BSIM3 model in a satisfactory manner;

2 The *bulk resistive network* is not included in the BSIM3 model;

3 The *physical gate resistance* is mostly not included in the set of extracted model parameters;

4 The *physical source/drain resistance* is mostly not included in the extracted model. Moreover the source/drain resistance parameters in the BSIM3 model[2] are only used in the I/V (current/voltage) model of the channel. Thus, they only represent a voltage drop within the channel but they do not introduce any additional poles in the AC simulations nor do they contribute any noise;

5 The *parasitic coupling capacitance* between the stack of metal fingers on top of the drain, source and bulk connection diffusion areas of the active devices is not fully included in the set of extracted parameters;

6 The *TML*[3]-coupling capacitance between the source and drain diffusion areas is not included in the model.

The first and second elements are shortcomings of the BSIM3 model itself, while the other elements are layout-related parasitics of the MOSFET device. These are difficult to include in the normal model files of the transistors because they are directly dependent of the physical layout the designer is drawing. Though not directly included in the MOST model files, the data to calculate these lacking parasitics is provided in the technology map, more specifically in the electrical design parameters. From these measured or simulated parameters, the designer can quite readily deduce the values of the additional parasitics. How this can be done is explained in the remainder of this section.

The first element that needs some further explanation is the one last mentioned in the list above, being the *TML*-coupling capacitance between the source and drain diffusion areas. Theoretically spoken, the fingers of source and drain diffusion can be regarded as planar transmission lines on a silicon substrate with plain air above them. Using the standard formulae for this kind of transmission line, the capacitive component of the coupling between the two lines can be calculated ([Wade 91]). As shown further, a comparison has been performed between a construction-based RF MOST model and a measurement-based model. To obtain a similar drain and source capacitance for both models,

---

[2]These parameters are *hdif* and *rsh*
[3]TransMission Line

the TML-component has to be added. This method has been applied here as an ad-hoc solution to construct a model in the case no measurement-based model is available. Since no further research has been performed to qualitatively or quantitatively verify the validity of this approach, it is not recommended to use it before measurements can really prove the presence of the TML-coupling. The fact that in measurements the effect of the TML-coupling is almost indistinguishable from the capacitive coupling between the drain and source metal fingers of the transistor, does certainly not raise hopes on a quick answer to the question whether this TML-effect really exists or not.

To implement all the other elements as depicted above, a subcircuit approach is used in accordance with [Enz ACD99] and [Leen 01]. From the MOSFET model as provided in the model files, an *intrinsic* MOST model is derived by nulling all layout-dependent elements. These include all series resistances in the model file, but also all parasitic junction capacitors. The latter are excluded, firstly because they depend on the physical layout of the transistor, secondly because the use of the "GEO"-option in Spice to indicate the use of a finger structure leads to an underestimation of the junction capacitances for a small number of fingers and finally because the implementation in BSIM3 does not allow for the addition of a bulk resistance network. Thus, an intrinsic model is obtained that only includes the channel model of the transistor and all parasitic capacitances of the gate. Starting from this intrinsic MOSFET, a parameterizable RF-MOST subcircuit is built, as shown in Fig. 2.4.

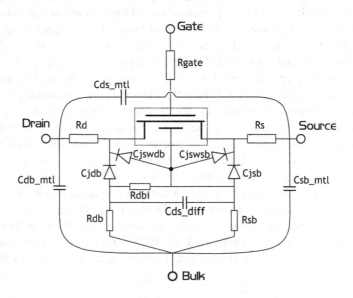

Figure 2.4.   The RF CMOS MOST model used in this work

Herein, the following elements are used:

Rgate: The physical resistance of the polysilicon gate and its contact vias;

Rs, Rd: The physical resistance of the diffusion in the source and drain areas and its contact vias;

Cjdb, Cjsb: The bottom-plate junction diodes between the drain resp. the source area and the well;

Cjswdb,Cjswsb: The sidewall junction diodes between the drain resp. source sidewall and the well area. For the sidewall at the edge of the transistor, this is uncharged well area, but for other sidewalls, this is the area underneath the conductive channel. This together with the fact that for sidewalls towards the channel an LDD extension of the diffusion area exists, explains why two different values for the sidewall capacitances are used. In BSIM3 the "normal" sidewall capacitance is given by *Cjsw* and the LDD sidewall capacitance is given by *Cjgate*;

Rdbi,Rdb,Rsb: The parasitic bulk resistors;

Cds_mtl, Cdb_mtl, Csb_mtl: The parasitic coupling capacitors between the stack of metal drain, source and bulk connect fingers. The coupling capacitance between source and drain is dependent on the channel length (L) of the MOSFET. The coupling between source or drain and bulk connection is dependent on the distance between the bulk connection and the MOST finger closest by;

Cds_diff: The coupling between the diffusion areas of drain and source, modeled as transmission lines. The presence of this effect has only been justified by a crude comparison of simulation results obtained using an ad-hoc constructed model and a measurement-based model. The validity of this approach should be further proven by measurements.

It can be observed that no parasitic coupling capacitances Cgs_mtl or Cds_mtl between the metal conductor terminals of respectively gate and drain and between gate and source are included in the model of Fig. 2.4. The reason for this is that they can be made negligibly small compared to the other capacitances if care is taken during layout.

From Fig. 2.4 a fully-scalable Spice subcircuit is derived. All layout information that has an influence on the parasitics of the subcircuit, is given as a parameter list with each instantiation of the subcircuit. The values of the parasitic elements are then calculated automatically. As an illustration, some lines of the Spice subcircuit are given in Code Ex. 2.1.

Code Example 2.1 Scalable Spice subcircuit for RF MOS device

```
*Subcircuits with scalable RF-MOST models

*Electrical parameters from electr. rules-map + layout rules*
.param  lmin=      0.24u
+       rsh_poly=    2.5  Poly sheet resist. [Ohm/sq]
...

*Junction diode parameters (from BSIM3v3 - model file )*
+       rsh_diffn= 3       BSIM3v3 "rsh"-parameter; the  n+
                           diffus. sheet resist. [Ohm/sq]
+       hdifn=     3.6e-7  BSIM3v3 "hdif"-parameter;
                           2*hdif= minimal diffusion width
+       cjn=       9.27E-04  BSIM3v3 "cj" for nMOS-model
...

*Subcircuit for scalable RF-version of nmos with SINGLE
  connected gate
.subckt rfnmos  dr gt sc bk wf=3.5u lf=lmin nf=4 ngt_cnt=1 ...
*Needed parameters from subcircuit:
*ngt_cnt: nu. of gate-contacts per finger to metal1
*wf_cnt : the space for extra diffusion contact
*nvia_src: nu. of vias connecting M1 to M2 for source
*nvia_dr:  nu. of vias connecting M1 to M2 for drain
...

.param  ndiff_cnt= 'int(wf/wf_cnt)'
+       Rgt_res1= '((1.0/3.0)*(wf/lf)*rsh_poly    Gate itself
                  + rcnt_gt/ngt_cnt)/nf'
+       Rgt_res2= '(gt_ext/lf)*rsh_poly/nf'  Gate connection
+       Rgt= 'Rgt_res1 + Rgt_res2'
+       Rsrc='(rvia/nvia_src+rcnt_diff/ndiff_cnt
              +rsh_diffn*(hdifn/wf))/nu_src'
...

*Implementation
Mi di gi si bi nint w=wf l=lf m=nf geo=0
Rsrcext si sc Rsrc      Source resistance
Rgtext gt_i gt Rgt      Gate resistance
...

*Mismatch on Vt and Beta
Vmmds dr dr_i 0
Vmmvt gt_i gi 'mm_nmos*delta_Vtn'              Vt mismatch
fmmbeta dr sc vmmds 'mm_nmos*reldelta_idsn'    Beta mismatch
```

As becomes clear from this code excerpt, the automated calculation of the parasitic elements is based on a set of equations and rules, some of which are found in literature and others are based on a detailed study of the physical layout of a MOSFET device combined with a study of the BSIM3 model. To give the reader a flavor of the knowledge built into the subcircuit model, the most important equations and rules are given here:

- The drain and source resistance are inverse proportional to the total drain respectively source length. The resistance per unit length is defined by the distance from the channel edge to the middle of the drain/source contact. The parameter *hdif* in BSIM3 quantifies this distance;

- The gate resistance that has to be taken into account for a single-side connected gate is smaller than the total physical resistance due to its distributed nature ([Raza TCASI94]):

$$Rgate = \frac{R_\square poly}{3} \cdot \frac{W_f}{L} \frac{1}{no_{finger}} \tag{2.1}$$

with $R_\square poly$ the sheet resistance of the polysilicon gate, W_f the finger width, L the finger length and no_{finger} the number of fingers. For a double-side connected gate the resistance is 4 times smaller [Tsiv 99];

- For the parasitic resistance calculation of both gate and S/D , the resistance of the contact vias is also taken into account;

- For an odd number of fingers, the number of source and the number of drain fingers is the same. For an even number of fingers, a choice must be made whether the source or the drain takes the extra finger, and its associated junction capacitance;

- The drain-bulk and source-bulk areas are modeled as junction diodes with their parameters taken from the BSIM3v3 model file. For both nMOST and pMOST three diode models are used, corresponding to the bottom plate junction, the diffusion-to-bulk sidewall junction and the LDD diffusion-to-channel sidewall junction;

- If wanted, even mismatch can be added to the model. This is done by assigning to the parameter *mm_nmos* the number of times the one-σ value of mismatch must be added to the transistor. The mismatch is modeled according to [Pelg JSSC89].

2.3.2.3 Model Validation

To validate the model, a comparison has been made in a $0.25\mu m$ CMOS technology between simulations done with the constructed subcircuit and simulations performed with an RF model provided by the foundry. The latter is a fit-model derived from measurements on a characterization batch of MOSFET devices with a limited size-range. The resulting scalable model is only valid within that range. Since the minimum finger length of the batch is $10\mu m$, the model turned out to be rather useless for real high-frequency design. This is briefly explained here:

Optimal MOST finger length   Simulations with varying finger length W_f in a $0.25\mu m$ CMOS technology show that a value between $3.5\mu m$ and $10\mu m$ is optimal for RF nMOS devices with a gate connected on one side. Devices with shorter finger lengths suffer from the higher series resistances in source and drain due to a lower total number of contact vias that can be placed, from the relative increase of the parasitic capacitance related to the gate extension over field oxide and from the relative increase of parasitic capacitance and resistance of the polysilicon used as finger interconnect. In devices with longer finger lengths, the influence of the gate resistance becomes dominant.

For the model validation, the simple common source nMOST amplifier of Fig. 2.5 has been simulated in AC. Since the lowest finger width available in the foundry model is $10\mu m$, the finger size for this comparison has been fixed at that lowest value. Two transistors with similar (W/L) but different channel length L are compared. The dimensions of the two transistors and their most important small-signal parameters for both the fit-model of the foundry as the construction-based model are given in Table 2.2. The values for the drain capacitance (*Cdtot*) are clearly larger in the fit-model than in the construction-based model. This is partially compensated for by adding the source-drain coupling capacitors *Cds_diff* and *Cds_mtl*. Although the predicted f_T-value for the transistor is somewhat pessimistic in the constructed model, the 3dB bandwidth of the amplifier (*BW*) is still overestimated by this model. However, the obtained accuracy of 5% is acceptable for circuit simulations.

In Fig. 2.6 the voltage transfer functions of the amplifier are shown; for the construction-based RF model drawn in full and for the fit-model from the silicon foundry drawn in dashed line. On the left a comparison is shown for transistors M1a and M1b of Table 2.2 with a gate length of $0.24\mu m$ and on the right for nMOS transistors M2a and M2b with a gate length of $0.5\mu m$. As can be seen, the construction-based model fits rather well to the measurement-based foundry model for frequencies up to 10 GHz for a gate length of $0.24\mu m$ and up to 6 GHz for a gate length of $0.5\mu m$. Of course, this comparison is far from complete since the use of the transistor in a totally different circuit could show shortcomings of the model that do not show up in this circuit. Moreover, the input impedance of the two transistor models has not been compared. However, since the MOS device will be used here in similar circuits as the test setup of Fig. 2.5, the model can be considered validated for the circuits described in this text. Hereby, the assumption is made that the construction-based model will also be valid for the lower finger widths that are commonly used in RF building blocks.

Table 2.2. Small-signal parameters for model validation

Transistor	M1a	M1b	M2a	M2b
Model	Constr.	Fit	Constr.	Fit
W_F	$10\mu m$			
N_F	11		21	
L_F	$0.24\mu m$		$0.5\mu m$	
(W/L)tot	458		420	
Rgate	4.44 Ω	5.67 Ω	1.36 Ω	4.28 Ω
Rd	0.8 Ω	-	0.44 Ω	-
Rs	0.25 Ω	-	0.14 Ω	-
Cds_diff	14.8 fF	-	28.3 fF	-
Cds_mtl	8.6 fF	-	10.8 fF	-
Cgtot	236 fF	165 fF	725 fF	545 fF
Cdtot	123 fF	135.5 fF	228 fF	249.6 fF
I_{ds}	4 mA			
Vth	0.508 V	0.535 V	0.522 V	0.574 V
Vgs-Vt	0.263 V	0.28 V	0.287 V	0.302 V
g_m	23.7 mS	22.8 mS	23.87 mS	23.02 mS
f_T	16 GHz	22 GHz	5.2 GHz	6.7 GHz
A_{v0}	9.32 dB	9.16 dB	10.26 dB	10.24 dB
BW	5.04 GHz	4.81 GHz	3.11 GHz	2.86 GHz

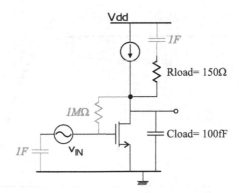

Figure 2.5. Common source nMOST amplifier used for RF-model validation

The non-quasi-static effect (in short: *NQE*) is not included in these models, since at the frequencies and for the applications used in this work, it does not have a big influence on the circuit behavior. However, since the importance will become bigger for the newest CMOS technologies and applications working in the multi- GHz range, a short description is given in the next paragraph.

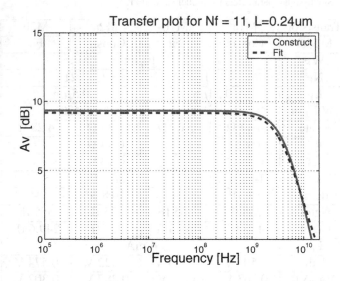

(a) 11 fingers, L=0.24μm

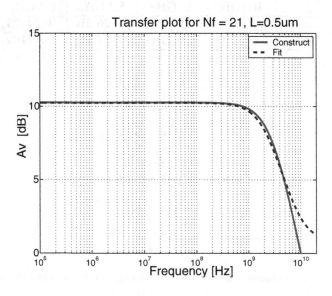

(b) 21 fingers, L=0.5μm

Figure 2.6.   Measurement-based vs. construction-based RF-model

2.3.2.4 The Non-Quasi-Static Effect

The standard I/V description of the channel of a MOSFET device is based on a few assumptions. Underneath the gate of a MOSFET, a channel charge is built up that has a certain distribution dependent on the gate, source and drain voltages. One of the most important assumptions made is the so called *quasi-static* assumption. It is assumed that the charge distribution in the channel has a profile that instantly adapts itself to any changes in the voltages it is dependent on. This is also expressed as:

> The time constant of the charge distribution in the channel of a MOSFET device is negligible compared to other time constants influencing the device behavior.

Since this time constant has a value in the order of 1 to 10 psec, this statement clearly seems to make sense for frequencies up to 10 GHz. However, in some specific topologies like LNA's, the non-quasi-static effect becomes visible at much lower frequencies than what can be expected looking solely at the time constant of the NQE. To understand this, the influence of the NQE on the observable MOS behavior is further investigated by giving a description of the effect and a short overview of existing approaches to model the NQE.

As any time constant in a system does, the time constant of the charge distribution introduces some delay. In this case, this means a time delay between a variation in v_{gs} and the moment a new equilibrium state in the channel's charge distribution has been established[4]. Let's suppose the time constant of the charge distribution in the gate itself is negligible as compared to the time constant in the channel. Then the charge distribution in the gate is in balance with the charge distribution in the channel at every instant of time. The time delay τ_{NQE} the build-up of charge in the channel experiences then leads to an identical delay in charge build-up in the gate. Thus, the (AC) gate current will be smaller than expected, and this behavior is observed as being similar to the case there were a series resistance $R_{gate,NQE}$ present in the gate. The channel *current* i_{ds}, on its turn, will flow with a time delay τ_1 in comparison with $g_m v_{gs}$[5]. This time delay is in first order twice as large as the time delay τ_{NQE} the *charge* in the channel experiences, since the current flow suffers from both the time delay in the channel and the induced time delay in the gate ([Tsiv 99]). The effect of time delay in the channel current can be modeled in first order by replacing g_m with $g_m \cdot (1 - j\omega\tau_1)$. This can be rewritten using the transcapacitance C_m of the channel, defined as the derivative of the channel charge to the gate voltage ([Enz ACD99]):

$$i_{ds} = Y_m \cdot v_{gs} = (g_m - j\omega C_m) \cdot v_{gs} \qquad (2.2)$$

---

[4]In fact, this delay exists for every terminal of the MOST that has an influence on the charge distribution in the channel, but this is not elaborated here.

[5]This time delay can also be expressed as a (frequency dependent) phase difference.

with Y_m the transadmittance between the drain-source port and the gate-source port of the MOS device. Since charge-based compact models as BSIM3v3, EKV and MOS Model 9 are inherently based on transcapacitances, this delay or phase shift in the channel current is present in simulations based on these models. However, although the delay is modeled correctly, the magnitude of the transadmittance as defined in (2.2) increases with frequency whereas it should be decreasing. A correct first-order model incorporating both delay and input resistance can be constructed by adding a VCCS to the basic transistor model and using a non-charge-based capacitance model to avoid the problem of the increasing admittance magnitude of (2.2) ([Enz ACD99]). In BSIM3v3, a similar solution has been provided ([BSIM3]) albeit at the cost of longer simulation times and possible convergence problems.

A simplification often used for the design of RF circuits operating at "normal" frequencies for which $\omega\tau_1 \ll 1$, boils down to the mere addition of a resistance to the gate node in series with the already mentioned physical gate resistance (2.1). The value of the additional resistance is given by [Leen 01]:

$$R_{gate,NQE} = \frac{1}{\kappa g_m} \tag{2.3}$$

which is valid for saturation, with $\kappa \approx 5$ - 7. The resulting model in fact neglects the time delay τ_1 of i_{ds}, without neglecting the effect of the charge delay τ_{NQE} on the perceived gate impedance. This approach is equivalent with a first-order modeling of the measured input impedance of a MOSFET gate.

The value of the gate resistance as given in (2.3) can be interpreted as a *fraction* of the total channel resistance, because of the distributed nature of the channel. Omitting the NQE is in fact equivalent with the representation of the distributed channel as a lumped element. The physical MOST channel is represented in the compact models by a single "node" that contains a certain amount of charge and has a well-described I/V behavior. An often used approach to improve the model accuracy when dealing with a lumped representation of a distributed element, is to divide the element in segments that are small enough for again a lumped-segment description to hold. Generally, the resulting global model containing those segments will more accurately describe the distributed behavior. Since all channel models for MOSFETs are extracted for a non-segmented device, this is not a straightforward approach, though not impossible [Tiem ESSDERC99]. In [Ou ACD99], the segmentation approach is used in an analytical way to calculate the value of the gate resistance due to the NQE. In essence, this boils down to the same result as depicted above in (2.3).

Now one question remains: when do we need to include the NQE in our circuit simulations? Two aspects of the NQE must be separated: the time delay τ_1 in the current and the equivalent gate input resistance $R_{gate,NQE}$ with its associated noise contribution.

- To decide whether the time delay is negligible or not, the transit time of the channel has to be compared to the rise time of the waveform that travels through the channel.

The transit time of the channel is derived in [Tsiv 99]:

$$\tau_0 \propto \frac{L^2}{\mu(V_{GS} - V_T)} \tag{2.4}$$

As a rough rule of thumb, if the rise time τ_R of the waveform satisfies

$$\tau_R > 20\tau_0 \tag{2.5}$$

then the delay introduced by the NQE can be neglected. This can be interpreted as the channel having an electrical length that can be regarded as "short".

In fact, this requirement is merely another way to demand that $\omega\tau_1 \ll 1$, but with an explicit indication of the dependency of τ_1 on the channel length and on the mobility of the majority charge carriers in the channel. For minimal length nMOST devices, the maximum allowable frequency is in the order of tens of GHz. For pMOS devices, the frequency at which the NQE becomes important is lower since τ_1 is larger for a pMOST due to the lower mobility of holes in silicon. Therefore, non-minimal length pMOST transistors may suffer from NQE even in "normal" RF applications with frequencies of a few GHz. This may necessitate the use of an elaborate model for the NQE, instead of the simple model represented by (2.3);

- For some RF building blocks, such as LNA's, also at lower frequencies the non-quasi static effect can have an impact on the performance. This is demonstrated in measurements of the input impedance of an LNA in [Jans ACD98], where the gate capacitance of the input MOST is tuned out by a series inductor. The resistive input impedance is clearly identified as coming from the NQE. Therefore, the design of such circuits must take this resistance into account to optimize their performance [Jans EL99];

- The noise contribution of the NQE is associated with the NQE resistance mentioned above. One could expect the noise contribution to be simply $4kT \cdot R_{gate,NQE}$. However, detailed calculations show that the NQE has a noise contribution given by ([Tsiv 99]):

$$S_{v,NQE} = \frac{4}{3} \cdot (4kT \cdot R_{gate,NQE}) \tag{2.6}$$

in which $S_{v,NQE}$ is the noise spectral density of a voltage source added in series with the gate resistor.

In this work, the effect of the NQE resistance has been evaluated in the circuits that are designed, and it has been found to be negligible as compared to the physical gate resistance. Since the inclusion would call for an additional iteration in the simulation process because $R_{gate,NQE}$ is inverse proportional to the value of g_m (2.3), it is not included in the RF model described in Fig. 2.4 altogether.

2.3.2.5 Frequency-Performance Indicators

A brief description is given of a few indicators that are used to quantify the frequency performance of MOS devices. The cut-off frequency f_T (also called *Transit* frequency) of a MOS device is the frequency at which the small-signal *current* gain drops to unity. This frequency is calculated by regarding the transistor as a two-port, represented by the hybrid 2 two-port model [Chua 87]. For a current-only, thus zero-voltage output the h_{21} parameter directly relates output current to input current. Setting it equal to one gives us the unity-gain frequency for the current. For a MOS device it is given by:

$$f_T \cong \frac{g_m}{2\pi C_{gtot}} \tag{2.7}$$

with C_{gtot} the total capacitance at the gate of the device, including all layout parasitics. Obviously, the value of g_m is proportional to the $V_{gs} - V_t$ of the device. Consequently, the f_T of a MOS device is expected to scale as $1/L_{eff}^2$, but this is only true in the absence of velocity saturation. If the carriers have a maximum velocity v_{sat}, the maximum obtainable transit frequency for the technology is given by:

$$f_{T-max} \cong \frac{v_{sat}}{2\pi \cdot L_{eff}} \tag{2.8}$$

The maximum transit frequency in the region of velocity saturation thus scales with $1/L_{eff}$. Since in weak and moderate inversion the frequency still scales with $1/L_{eff}^2$ because velocity saturation is less likely to occur, the migration towards deep submicron technologies might enable the use of devices biased in the region of weak and moderate inversion in high-speed applications.

The expression of (2.7) does not take any parasitic resistances into account and is therefore not useful to quantify the performance of a certain technology with respect to the design of RF circuits. A better number to represent what frequencies a certain technology can handle, is the *maximum oscillation frequency* defined as the frequency for which the unilateral *power* gain of the transistor becomes unity. Above this frequency, the total loss in the parasitics of the device becomes larger than the power gained by using the transistor as power amplifier. This figure thus encompasses both current and voltage gain, and includes all parasitics of a single device. Different expressions for this frequency exist, of which two are given here. The first one [Enz ACD99] relates f_{max} directly to the physical gate resistance R_{gate}:

$$f_{max} \cong \sqrt{\frac{f_T}{8\pi R_{gate}C_{gd}}} \qquad (2.9)$$

The second one [Leen 01] also includes the effect of the bulk resistance and the source/drain resistance:

$$f_{max} \cong \sqrt{\frac{f_T}{4\pi C_d \sqrt{(R_{gate} + R_{gate,NQE} + R_s)R_{\text{eff}}}}} \qquad (2.10)$$

with C_d the junction capacitance of the drain, R_s the source resistance and R_{eff} the effective bulk resistance. In general, the value of f_{max} will be higher than that of f_T.

2.3.2.6 Summary

A scalable model for a MOS device has been constructed based on available data of the technology provider and on a detailed study of the layout of a MOS transistor. The model is implemented as a Spice subcircuit for use in circuit simulation. A short description of the non-quasi-static effect is given and it has been indicated when it has to be taken into account. In the circuits designed in this work, the NQE can be omitted in simulations. In a last part, the major frequency-performance indicators for MOS transistors are summarized.

2.3.3   On-chip Inductors

2.3.3.1   Introduction

Over the past few years, the use of on-chip inductors has had a growing interest, both in academic research as in industrial applications ([Sama ISSCC01, Fili ISSCC01, Su ISSCC02, Apar ISSCC02]). An on-chip inductor is used as an alternative for ([Cran CICC97, Crols VLSI96]):

1   a bondwire inductor

2   an off-chip inductor, being a discrete component

The advantage of using an on-chip inductor is the higher reproducibility of the inductance value as compared to (1) and a lower assembly cost combined with a higher reliability due to a reduced number of discrete components as compared to (2). The major drawback of on-chip inductors generally is the low Q-factor and the large chip area used. As demonstrated already in Section 2.2.3, the problem of the low Q-factor can be alleviated by using a technology with a high-resistivity substrate. As mentioned in Section 1, the relative large area of an on-chip inductor might cause it to be more economical to use an older, mature CMOS technology instead of the newest deep-submicron technology.

One of the key issues in the use of an on-chip inductor is a reliable prediction of its behavior. A widespread approach is the straightforward method of producing and measuring a whole batch of inductors with varying size and/or geometry. The measurements lead to a fully-characterized library of inductors that can be used by the chip designers. In some cases, a model is fit to this measured batch. Using this approach, the available set of inductor sizes is restricted to a certain subspace of all available inductor sizes. The larger the amount of inductors in the batch, the higher the variety of available inductors in the library becomes or the more general and/or accurate the derived inductor model will be. Major drawbacks are the high area cost of the inductor batch and the time delay introduced by the characterization. The higher the number of inductors in the test batch, the more clearly this drawback will manifest itself.

Another approach to the design of on-chip inductors is also possible. Some kind of inductor simulator can be used to predict the behavior of the on-chip inductor without a fabrication and characterization step prior to the design step. In this work, this method is adhered to. As a tool for the simulation of the inductors, the freeware package FastHenry is used [Kamo DAC93].

In a first section a short overview of the high-frequency effects in on-chip inductors is given. The second section gives a non-exhaustive overview of available techniques for inductor parameter prediction. In the third section, FastHenry is introduced and the basic working principles are explained. FastHenry is used to derive the inductance and resistance value of the inductor at one specific frequency. The last section describes how circuit models are derived from the FastHenry simulation.

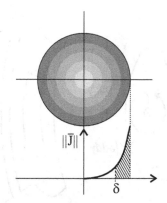

Figure 2.7. Visualization of the skin effect

2.3.3.2 High-Frequency Effects

A very short overview of the effects that influence the behavior of an inductor at high frequencies is given here [Cran PhD97]. It will be used to clarify the differences between the alternatives for inductor parameter prediction as described in the next subsection.

Metal losses Two frequency dependent effects influence the metal losses. The first effect that comes into play at high frequencies is the *skin effect*. Due to the self inductance of a conducting wire, the current flowing through it is pushed to the outside of the wire. As visualized in Fig. 2.7, the skin depth δ_{skin} is defined as the distance from the outer edge of the conductor at which the current density is reduced to a value of $1/e$ compared to the current density at the edge. For a metal-like conductor, this skin depth is given by:

$$\delta_{skin} = \sqrt{\frac{2}{\mu\sigma\omega}} \qquad (2.11)$$

The higher the frequency, the lower the skin depth thus the higher the (local) current density in the conductor. The effective resistance the current is experiencing therefore rises with frequency.

The second effect is more specific to an on-chip planar coil. Fig. 2.8 shows how the alternating current I_{coil} flowing through the turns of the coil generates an alternating magnetic field $\vec{B}_{coil}$. This field at its turn generates eddy currents that rotate within the turns of the coil in a direction indicated by I_{eddy}. The magnetic field $\vec{B}_{eddy}$ counters the magnetic field that generated the eddy current. I_{eddy} causes the current flow in the inner turns to be concentrated at the inside of the turn. The higher current density here again leads to a higher resistance. $\vec{B}_{eddy}$ causes a partial suppression of the global magnetic field, resulting in a lower coil

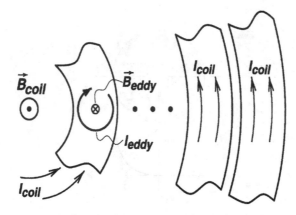

Figure 2.8.   Eddy currents in planar inductors

inductance. Due to the higher magnetic field density in the middle of the coil, the effects are more pronounced in the inner turns, and since the magnitude of the induced magnetic fields and currents is proportional to the frequency, also the rise in resistance and the loss of inductance will be proportional to the frequency.

Substrate loss   The use of a standard CMOS technology, prohibits the use of post-processing steps like wafer etching to remove the substrate material underneath the integrated coil. Therefore, the presence of a conducting substrate with its deteriorating influence will always have to be taken into account. In Fig. 2.9 a cross-section of an on-chip coil is shown. The magnetic field $\vec{B}$ is induced by the alternating current I flowing through the coil. This magnetic field also alternates, and similarly to the effect on the metal of the coil itself, it also induces eddy currents I_{ind} in the substrate. On the one hand, these currents suffer from losses in the substrate; on the other hand they induce a magnetic field that tries to counter the originating field $\vec{B}$. The power loss in the substrate results in a higher effective resistance of the coil, and the partial suppression of the magnetic field results in a lower inductance. As already demonstrated in Section 2.2.3 a high substrate resistivity is beneficiary for the performance of RF building blocks with on-chip inductors. This can be explained based on this substrate loss. In a low-resistivity substrate, a larger current flow occurs, resulting in a more substantial power loss than in a substrate with a higher resistivity. The additional power loss due to the substrate coupling of the magnetic field results in a lower Q-factor of the on-chip inductor. The introduction of a patterned substrate shield to avoid the eddy currents from flowing only partially resolves the problem since the magnetic field penetrates

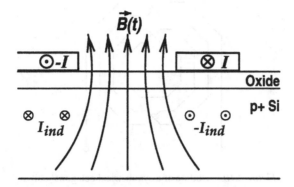

Figure 2.9. Induced substrate currents underneath a planar coil

through the whole substrate whereas the shield pattern is only present in the top. The improvement that was nevertheless demonstrated in designs [Yue DAC99] only emphasizes the high impact of the substrate loss on the performance of on-chip inductors.

Conclusion From the description of the effects that influence the behavior of planar on-chip inductors, it becomes clear that the substrate effects need special attention during the design of these components. The next section gives an overview of the different existing methods for inductor parameter prediction. The accuracy of those predictions will be directly linked to the way the substrate effects are taken into account (or not).

2.3.3.3 Inductor Parameter Prediction

Different methods exist to predict the parameter values of an on-chip inductor. Of these, probably the most reliable is the use of measurements but it comes with the penalty of a serious time delay needed for chip processing and measurements. In Table 2.3 some of the existing methods for parameter value prediction are compared. A good property is indicated by a high number of "+", a bad property by a high number of "-". The properties used to compare these methods need a short explanation. In the descriptive list given below, the abbreviations used in the table are given in between brackets:

Pre-Use Time *(PUT):* This is the time needed before the method can even be used to start making predictions. For measurement-based methods this includes the time for processing and measuring a batch of inductors; for geometry-based simulation methods this could be the time needed to write an automated geometry file generator adapted to the input format of the simulator;

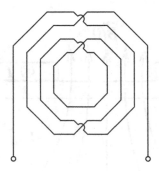

Figure 2.10.   Outline of a symmetrical octagonal coil

Set-Up Time *(SUT)*: The time needed to prepare the input of the prediction method for a single inductor before the evaluation itself starts;

Evaluation Time *(ET)*: This is the time needed to obtain the parameter values of a single inductor;

Flexibility of use and implementation*(FLEX)*: The relative ease of adaptation to slightly different geometries, and also the ease of use within an automated optimization algorithm;

Symmetrical octagonal coil *(S8)*: An "X" means that this method can also predict the parameter values for a *symmetrical* octagonal coil as shown in Fig. 2.10, an "O" means that it can not and a "?" means that it has not been explicitly shown that it is possible;

Inductor/Resistance Value Accuracy *(I/RVA)*: In the discussion per method given further on, the reported percentage of accuracy obtained is given, if available;

Cost: The cost of the method. In between brackets it is indicated whether this cost is recurrent (R) for each change in technology or independent (I) of the used technology.

For the different methods depicted in Table 2.3, the relative evaluation is explained in detail in the overview given here:

Inductor library, no model: The *PUT* consists of the layout, fabrication and measurement of the batch of inductors that form the inductor library. Since for each inductor available to the designer the exact values of all parameters are known, this approach has quasi-zero *SUT* and *ET*. However, the flexibility is very low since inductor geometries not present in the library simply cannot be used, and optimization is limited to a try-and-error technique using the available library. In general, the accuracy of the parameters

Table 2.3. Methods for inductor parameter prediction

	PUT	SUT	ET	FLEX	S8	I/RVA	Cost
Inductor library *no model*	$--$	$++$	$+++$	$--$	X	$+++$	$--$(R)
Inductor model *meas. based*	$---$	$+$	$+$	$-$	X	$++(+)$	$--$(R)
Closed-form eqn. *fitted*	$-$	$+$	$++$	$+$	O	$-^a/--$	$++$(I)
Closed-form eqn. *physics-based*	$++$	$+$	$++$	$+(+)$	O	$++/?$	$++$(I)
Inductor simul. *Greenhouse*	$-$	$+/-^b$	$+(+)$	$++$	?	$+/-$	$++$(I)
Inductor extract.	$-$	$+$	$- \rightarrow +$	$+++$	X	$++$	$++$(I)
Finite element	$-$	$+$	$-- \rightarrow -$	$++(+)$	X	$++(+)$	$-$(I)

---

[a]"+" for non-octagonal coils in model of [Mohan JSSC99]
[b]see text for explanation

is dependent on the extraction method, but once this is under control, excellent accuracy is achievable. For every change in the technology, the inductor characterization has to be done again, resulting in a recurrence of the high chip processing cost.

Inductor model, measurement based: This approach is a further evolution of the previous one, resulting in an even longer *PUT* due to the time needed to develop a good, scalable model. The *SUT* and *ET* are a little higher than for the previous method due to the need for implementation of the inductor model as integrated subcircuit within the total circuit and the time needed for the simulation itself. The flexibility is higher due to the scalability of the model and the possibility to use the model within an optimization loop. If only a spice subcircuit implementation of the model is available, possibly a circuit evaluation will be needed for each run of the optimization loop. The accuracy of the model itself will always be lower than that of a single measured inductor[6]. The cost is only marginally higher than the cost of the inductor library itself. For example silicon foundries as UMC and TSMC provide under NDA a scalable inductor model, both as closed form

---

[6]The obtained accuracy depends on the used fitting method. A Least-Square fit will minimize the global error over the whole range of the model, whereas other methods may minimize the local error at the available data points, possibly at the cost of a larger error on some other points within the modeling range.

expression and as Spice subcircuit. A drawback is that these models are seldom available for the early analog runs of a new technology.

Closed-form Equation, fitted formulae: The *PUT* is very short, since this kind of formulae are ready to use. In the derivation of the formula, a lot of CPU time is spent at the simulation and fitting of the proposed formula, but this is of no importance for the end-user. [Crols VLSI96] uses MagNet ([Free 93]) as finite element simulator and [Mohan JSSC99] uses ASITIC ([Nikn JSSC98]) to do the simulations. The *SUT* is comparable to that of the other methods. Here the geometry parameters of the inductor have to be recalculated into the parameters of the formula(e) as used by the method. The *ET* is short, and the flexibility is only slightly lower than that of the methods to follow. Although this kind of formulae is very well suited to be used within an optimization loop, it still has the minor drawback that a certain inductor shape needs a certain fit-formula. This kind of equations generally confines to predicting the *inductance* value of the inductor only. The accuracy of this prediction varies from acceptable to bad. Article [Jenei JSSC02] reports an accuracy of better than 6% for 80% of a variety of square inductors, but only 10% for 80% of a set of octagonal inductors for the monomial formula of [Mohan JSSC99]. In the same article, the use of the formula depicted in [Crols VLSI96] results in an accuracy of 12% for 80% of a set of square inductors. [Cran PhD97] reports for the latter formula also an indicative accuracy of the calculated series resistance of about 30-40%. This is clearly insufficient if the series resistance has any influence on the performance of the building block the inductor is used in. These methods use freely available equations.

Closed-form Equation, physics based: The equation can be used directly by evaluating it for the geometry parameters of the inductor, and the evaluation time is short, leading to a good *PUT*, *SUT* and *ET*. The flexibility is higher than for the previous method since one formula can be used to handle square and octagonal inductors and even inductors with uncompleted turns ([Jenei JSSC02]), while the same ease of use within an optimization loop is obtained as for the previous method. The inductance is calculated with an accuracy of better than 3% for 80% of a variety of inductors. As is also the case in the previous method the resistance is not or not accurately modeled, rendering these two methods useless for those RF designs where a good prediction of the resistance or Q-factor is wanted. The cost is low, since the equation is available for free.

Inductor simulation, based on the Greenhouse formula: Some time (*PUT*) is needed to learn to use this kind of simulators, and possibly to write a script for the automated generation of the used geometry. For programs like

ASITIC, [Nikn JSSC98] a technology-dependent file describing the electrical parameters and the layer construction of the technology has to be set up. The *SUT* is good if a script-based geometry file generation is used to supply the input file to the simulator. In the case the inductor geometry is to be drawn manually, the *SUT* is a lot worse. The *ET* is rather short. The flexibility is good, since arbitrary inductor shapes can be simulated, but still lower than the following two methods since the use of these simulators within an optimization loop is not always straightforward. The accuracy of the calculated inductance is lower than for the previous method as demonstrated in [Most TCASII01] and [Leen ICCAD01] where measurements are compared to simulation values obtained with ASITIC [Nikn JSSC98]. Other simulators in this category are reported in [Kouts TCASII00] and [Long JSSC97], the latter showing very good modeling accuracy for square-shaped inductors. Not all of these programs are freeware, but the cost is rather low.

Inductor extraction: This method has the same *PUT* as the methods above, and it also needs a technology-dependent file describing the electrical parameters and layer construction. The FastHenry program ([Kamo DAC93]) allows the use of a script-based geometry file generation, resulting in a good *SUT*. The *ET* constantly drops due to the progress in available CPU speed. The flexibility is very good since arbitrarily shaped inductors can be simulated and the use of the program within an optimization loop is quite straightforward to implement. The accuracy is good, both for *IVA* as for *RVA*, since substrate effects are taken into account. The FastHenry program is freeware.

Finite element simulation: This method has a *PUT* that is quite comparable with the previous one. Also the *SUT* is almost the same, since the use of a script-based geometry file generation is possible. Here, both an increase in available CPU speed and the research towards more time- and memory-efficient finite element algorithms [ELEN] explains the evolution from a bad to a reasonable *ET*. The *FLEX* is identical to that of the previous method, but the resulting *IVA* and *RVA* are almost as good as those obtained with direct measurements. Programs as [Free 93], implementing this method, are seldom cheap but the investment is not recurring for each new technology, explaining the indicated *Cost*.

Final remarks This subsection gives a non-exhaustive overview of available methods to predict the parameter values of on-chip inductors. The previous subsection already mentioned the influence of substrate effects on these parameters, in particular the effective series resistance of the inductor. A good correlation can be seen between the way substrate effects are taken into account and the obtained accuracy of the predicted series resistance, indicated as *RVA* in Table 2.3. The less accurate the substrate effects are modeled with the method, the less accurate the series resistance can be predicted. This first conclusion might seem obvious, but its importance is not to be underestimated; for industrial applications the accuracy of these methods simply is too low to be acceptable [Leen ICCAD01].

The *S8*-column in Table 2.3 is still uncommented up till now. The use of symmetrical octagonal coils instead of square coils has a proven advantage[7] in integrated VCO's what phase noise performance is concerned [Most TCASII01, Cran PhD97]. Therefore, only this type of inductors is used in this work. However, not every method foresees the modeling of this type of inductors. Especially the symmetrical construction of the coil, as shown in Fig. 2.10 and Fig. 2.11 seems problematic. Put simple, the methods we can choose from for the parameter prediction of inductors used in this work are those with an "X" in the *S8*-column.

Within an industrial environment, the financial cost associated with the processing of an inductor batch will seldom be a real burden if compared to the cost of a whole project. On the other hand, the high accuracy *theoretically* obtainable with the finite element method will only be certified after chip processing if the program is used in a proper way. This encompasses a good knowledge of the program itself ánd a good translation of the physical reality into a modeled reality which is presented to the program. Moreover, industrial partnerships between design groups or companies and technology providers generally enable the former to send their inductor batch with one of the final characterization runs of a new technology. This way, the time delay normally involved when this method is used, can be alleviated. Finally, the maturity and the proven reliability of this method help to explain why many design groups and companies still adhere to the use of an inductor model based on measurements.

For non-industrial environments, these pre-release runs are seldom available. Moreover, the whole set-up including measurements and model extraction of an extensive inductor batch calls for an amount of knowledge and resources that are beyond the (financial) grasp of (small) research groups. Based on the combination of good accuracy and reasonable evaluation time, the only really practical method that thus remains, is the induction extraction method. The fact that it is freely available only speaks at its advantage. The next subsection will deal with the use of FastHenry, explaining in detail the syntax and program parameters that influence the final results of the inductor simulation.

---

[7]One of the obvious advantages is that the use of a symmetrical coil in a differential topology results in a self-oscillation frequency being twice as high as that of the same coil used in a single-ended topology [Nikn ESSCIRC99].

2.3.3.4 The Use of FastHenry

FastHenry is a three-dimensional inductance extraction program, that is distributed at no charge. It can be used to calculate the inductance and resistive losses of a structure consisting of conductors of complex shape. The influence of a conductive ground plane, underneath the structure, is also taken into account in this calculation. The usability of the program depends on the resistivity of the ground plane. For ground planes with a resistivity higher than 5 Ω.cm, reliable results can be obtained. For lower ground plane resistivities, the calculation algorithm of FastHenry is not accurate enough to be used.

FastHenry is used in this work to calculate the complex impedance of on-chip inductors at a certain frequency. The possibility of the program to derive an inductor model based on simulations at multiple frequencies, is not used. The main reason for this is that only a two-port model can be derived, whereas the symmetrical coils used here have three externally accessible nodes.

The simulation of an inductor using FastHenry starts with the definition of its geometry. The circular shape is the most optimal what the ratio of inductance/area is concerned. However, this shape is mostly not allowed by standard DRC rules. Often used is the square shape, since this is easy to draw by hand and does not offend standard DRC rules. As already mentioned in the previous subsection, only symmetrical octagonal coils are used in this work:

1 Because of their better approximation of the "ideal" circular shape, while still not offending any standard DRC rules;

2 Because of the advantages of their perfect symmetry, especially in the design of differential VCO's.

The construction-by-hand of an octagonal coil is a little more time-consuming, and therefore programs are used to construct the coil geometry both in FastHenry input format as in GDSII format. An example of a symmetrical coil geometry as visualized by FastHenry is depicted in Fig. 2.11. The picture clearly shows the segments of which the coil is constructed.

These segments interconnect each two nodes in a straight line and are used to describe the coil geometry in the input file of FastHenry. The width and thickness of the segments is set to a common (default) value for all segments, corresponding to the metal width of the coil and to the *total* metal thickness of the used metal layers. This total metal thickness is the distance between the bottom of the first used metal layer and the top of the last used metal layer in the coil, thus including the inter-metal oxides. Fig. 2.12(a) shows the quite standard layer construction of a CMOS technology. The total metal thickness can also be regarded as the thickness $t_{virtual}$ of a *virtual* metal layer, as indicated in Fig. 2.12(b) for the case metal layers 2 up to the top metal layer are used in the inductor. In this drawing, the oxide thickness between coil and substrate is

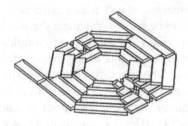

Figure 2.11.   Symmetrical coil geometry as simulated by FastHenry

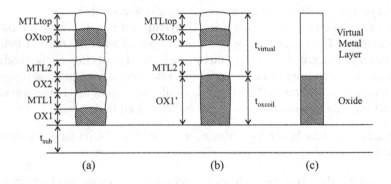

Figure 2.12.   Layer construction of a CMOS technology

indicated as t_{oxcoil}. The sheet resistances of each used metal layer, as found in the electrical rules map of the technology, are recalculated into a *virtual* conductance of the *virtual* metal layer indicated in Fig. 2.12(c). When enough vias are used in between the parallel metal layers of the coil, this simplification does not introduce any errors in the simulations. However, care must be taken for technologies with a low resistive, thick top metal layer (possibly in Cu) because for this non-homogeneous structure, the vertical current distribution might differ substantially between physical reality and simulation. The physical layer construction of the inductor as visualized in Fig. 2.12(b) is thus reduced to the simplified layer construction of Fig. 2.12(c), consisting of one virtual metal layer and one oxide layer.

The algorithm used in FastHenry divides the segments that are used to construct the coil geometry in smaller beams, called filaments, for a more accurate calculation of the field distribution in each segment. In Fig. 2.13 this discretization of a segment is visualized. Using this discretization, the current

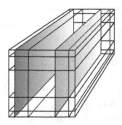

Figure 2.13.   Segment discretization as used by FastHenry

flow through the segments and through the inductor, is determined. As described above, high-frequency effects force the current in a conductor to be pushed away from the center. Therefore, the outer filaments should be smaller than the center filament to obtain a similar accuracy of the calculated field distribution over the whole segment. To accurately model all effects, the outer filament should have a width smaller than the skin depth at the simulated frequency. This has to be ensured by the user of FastHenry by setting the segment division parameters correctly, as explained further on.

In Code Ex. 2.2 the code of a typical input file for FastHenry is given. It consists of three parts: a file *header*, a middle part that defines the placement of nodes and segments and the file *footer*.

The code in the header sets the units to micron, and issues default values for some parameters of the coil segments:

- the value of the height z above the substrate is set at 5.85μm;

- *sigma* is the equivalent conductivity of the virtual metal layer;

- h is the thickness of the virtual metal layer;

- the parameters *nhinc*, *nwinc* and *rw* define how the segments are subdivided in filaments. The first and second parameter define the number of segments in respectively vertical and horizontal direction. The last parameter sets the ratio of two adjacent filaments

Discretization example: For rw= 2 and nwinc= 5, the sizes of the filaments in the horizontal direction have a relative size given by: 1 2 4 2 1, meaning that the middle filament is 4 times larger than the outer filament. The size of the thinnest, outer filaments is in this case $1/(1+2+4+2+1) = 1/10$ of the width of the wire.

Code Example 2.2 FastHenry input file

```
*Coil input file: Example

** Input File Header **
*Technology name: umc025_4T
*   UMC 0.25um CMOS (4 Top Layers used)
* Total Metal Thickness: 5.7 um
* Thick Oxide Thickness: 3.0 um
* Equivalent Metal Conductance: 15.0 1/(Ohm.um)
* Substrate Conductance: 5e-6 1/(Ohm.um)
* Epi Layer Thickness: 0 um

.units um
.default z=5.85 sigma=15.0 h=5.7 nhinc=3 nwinc=7 rw=2
** Input File Coil Description **
*symmetrical coil layout: w = 43.8, r = 196.0, turns = 3,
                leads =100.0
.default w=43.8

n3 x=-174.1 y=-174.1
n4 x=174.1 y=-174.1
n1 x=-174.1 y=72.1
n2 x=174.1 y=-72.1
e1 n3 n1
e2 n4 n2

...

.equiv Nout n65
.equiv Nin n0

** Input File Footer **
G1 x1=-1000 y1=-1000 z1= -362.5
+   x2=-1000 y2=1000  z2= -362.5
+   x3=1000  y3=1000  z3= -362.5
+   thick=725 sigma=5e-6 nhinc=5 rh=2 file=NONE
+   contact decay_rect (0,0,0,750.0,750.0,75,75,2000,2000)
.external Nin Nout
.freq fmin= 2200000000.0 fmax= 2200000000.0 ndec=1
.end
```

In general, the size of the outer filaments can be calculated from the sum of the relative sizes of the filaments. This sum of relative sizes is given here, as function of the FastHenry parameters rw and nwinc, for odd values of nwinc:

$$sum = 2.\frac{rw^{\frac{nwinc-1}{2}} - 1}{rw - 1} + rw^{\frac{nwinc-1}{2}} \qquad (2.12)$$

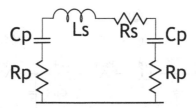

Figure 2.14. Simple inductor model

As already stated above, this size should be smaller than the skin depth of the used metal at the frequency of interest, as given by (2.11).

The coil described in the middle part of the file has a metal width of 43.8μm, an outer radius of 196μm and three turns. Two leads of 100μm to connect the coil with other circuitry are also part of the simulated structure. The width w of each segment is set to a default value of 43.8μm. The *n*-command is used to describe the position of the nodes of the structure, and the *e*-command is used to define a segment between two previously defined nodes. The *.equiv*-command makes nodes *Nout* and *n65*, respectively *Nin* and *n0* electrically equivalent.

The footer defines the ground plane by describing the position of three of its points. The thickness of the plane is set at 725μm, the conductivity *sigma* is equal to 5μS and the parameters *nhinc* and *rh* define the subdivision of the segments of the ground plane. These segments are situated within a mesh defined on the ground plane by the *contact*-command. Nodes *Nin* and *Nout* are selected as the external nodes of the coil. The simulation is performed at one single frequency of 2.2 GHz.

This input file is processed by FastHenry and the impedance between the two external nodes of the described inductor is calculated at the wanted frequency. From the resulting complex impedance Z_{FH} two different inductor models can be derived. This is dealt with in the next section.

2.3.3.5   Inductor Subcircuit for Circuit Simulation

Two different inductor models are used to represent the simulated inductor:

1 A simple model, consisting of the two elements Ls and Rs, being the inductance value and the total series resistance of the coil, including all high-frequency and substrate coupling effects. This two-element model can be further extended to the well-known six-element model of Fig. 2.14 by adding calculated values of the substrate coupling capacitance and its series resistance. This model is valid only at the simulated frequency, since the frequency dependence of the value of Ls and Rs is not included in the model;

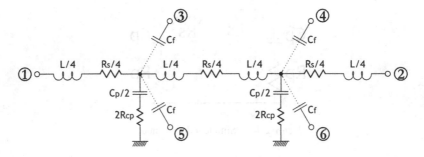

Figure 2.15.   Subcircuit for one half turn of a coil

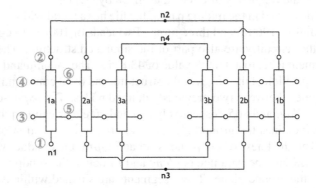

Figure 2.16.   Complete model of an on-chip coil

2 A distributed model, built up using fifteen-element subcircuits for each half turn of the coil. This approach is used in Chapter 3 to construct a Spice simulation model for the whole inductor that is also valid at frequencies differing from the FastHenry simulation frequency [Cran PhD97]. The subcircuit of one half turn is shown in Fig. 2.15 and like the simple model it combines values derived from the simulated impedance with calculated values for capacitances and their series resistances. A visualization of an inductor model for a coil with three turns is given in Fig. 2.16.

   For both models FastHenry is used to take the magnetic substrate coupling of the inductor into account and manual calculations are used to calculate the values of the extra subcircuit elements that take the capacitive substrate coupling into account.

   In the construction of the simple six-element model, care must be taken in the assignment of specific values to the different model elements. Since we use only symmetrical coils in this work, this is the only topology we will explain this for.

Assigning values to the simple coil model The easy part of the model construction consists of the assignment of values for Ls and Rs. Rs is simply the real part of the complex impedance Z_{FH} as returned by FastHenry. Ls is calculated as $Im(Z_{FH})/\omega$.

The total coupling capacitance Cp_{tot} between coil and substrate, consists of the bottom plate oxide capacitance and the sidewall stray capacitance of the coil. The series resistance Rp_{tot} between this capacitance and physical ground is calculated using the substrate resistivity and the mean distance from the coil to the nearest ground connection.

Since the central tap of the symmetrical coil is at AC ground, the voltage swing over the coupling capacitance gradually decreases from either input to the central tap. This is illustrated by Fig. 2.17 where the simulated voltages at different nodes of Fig. 2.16 are shown for the coil specified in Table 3.5. The path along the coil is traversed from node *n1* to *n4* for values of the *relative position along the coil* in Fig. 2.17 going from 0 to 40. Although the central node (*n4*) is not at an ideal AC ground, it is nevertheless clear that the voltage swing drops almost linearly from node *n1* to the central node. As any capacitance value, the capacitance seen at node *n1* is defined based on the total displaced charge for a certain voltage swing at *n1*. Since the voltage swing drops linearly, the total charge displacement through *n1* is calculated as the area underneath a triangle. Therefore, the physical capacitance as calculated above only has to be taken into account for 2/3 of its value due to the symmetrical use of the coil. Furthermore, the total capacitance and resistance of the model depicted in Fig. 2.14 is the parallel connection of the elements present. Consequently, the two elements Cp are both assigned a value of $\frac{1}{2} \cdot (2/3 Cp_{tot})$ and the value of the elements Rp is equal to $2 \cdot Rp_{tot}$.

The value assignment to the different half turns of the distributed coil model is done proportionally to the relative area of each half turn. Thus, the outer half turns will have a larger inductance, resistance and capacitance value. This way of assignment expands the usable range of frequencies for this model towards lower frequencies than the FastHenry simulation is performed at. At higher frequencies, the resistive contribution of the inner turns will become more important due to the relative larger eddy currents present in the inner turns, and this is not included in the model.

The whole process of model generation, both for the simple and distributed model, can be completely automated, and this will be used in a tool for the automated design of LC-type VCO's, as described in Chapter 3.

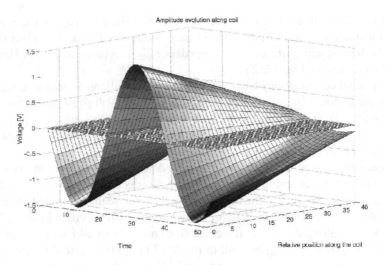

Figure 2.17.   Voltage swing along symmetrical coil of Table 3.5

2.3.4   Conclusion

With the shift towards ever higher frequencies, the importance of good models for the design of first-time-right silicon is becoming higher and higher. In this section the construction of a subcircuit model for a MOSFET device has been described. It is based on similar approaches found in literature and a detailed study of the physical layout of the device. The scalable subcircuit has been used for the design of an upconvertor circuit presented in Chapter 5.

Compared to other methods to predict the behavior of on-chip inductors, the freeware package FastHenry prevails because of its accuracy and ease of implementation within an optimization program as implemented in Chapter 3. Based on the simulated complex impedance at a single frequency, a lumped inductor model is constructed for circuit simulation. Due to the use of a multi-element subcircuit, the validity of the inductor model is extended towards lower frequencies.

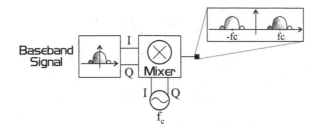

Figure 2.18. Quadrature upconversion

2.4 Quadrature LO Generation

Starting from a differential LO-signal, two different approaches have proven their reliability to generate on-chip a differential quadrature signal. The first one is the use of a polyphase network. A second approach to generate a quadrature signal is to use a divide-by-two circuit, necessitating an oscillator working at twice the desired LO-frequency. These well-known approaches will be briefly presented focussing on some issues that can have an influence on the performance of the two building blocks.

In the first section, the need for a quadrature LO-signal with low quadrature error for use in telecommunication circuits is explained. For a polyphase filter the influence of mismatch and of higher order harmonics of the local oscillator are discussed, and some practical design considerations are given in a second section. The third section deals with the divider-type quadrature generator.

2.4.1 Quadrature Upconversion

In Fig. 2.18 the schematic of a general quadrature direct upconversion mixer is shown. The goal of this kind of circuits is to generate a single-sideband signal at a certain carrier frequency, indicated as f_c in the drawing, starting from a quadrature oscillator signal at the carrier frequency and a quadrature input signal carrying the information at baseband frequency. [8]

In modern digital telecommunications systems, the baseband signal to be transmitted is in fact a digital bit stream. To increase the overall bandwidth efficiency this bit stream is first modulated using e.g. Quadrature Amplitude Modulation (QAM), of which a very simple modulation mask represented in the complex plane is shown in Fig. 2.19. Using the represented map (QAM4/QPSK), four different symbols can be transmitted after modulation into an I/Q pair. Herein, the I-component is the projection of the complex symbol vector $\vec{V}$

---

[8]Representing the in-phase component of the quadrature signal as the I-signal and the quadrature-phase component as the Q-signal, a quadrature signal is often also called an I/Q signal.

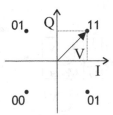

Figure 2.19. Modulation mask for QAM4/QPSK

on the real axis and the Q-component is the projection on the imaginary axis. Using this simple modulation scheme, the system is able to transmit 2 bits of information per symbol, if no overhead for error correction is used. This can be expressed mathematically as follows: the signal vector $\vec{V}[k]$ that represents a symbol of the symbol sequence $\{S[k], k = 1..n\}$, is modulated into two signals $I[k]$ and $Q[k]$ such that:

$$\vec{V}[k] = I[k] + j \cdot Q[k] \tag{2.13}$$

This is an expression consisting of discrete (digital) variables in the discrete complex time domain. The expression of (2.13) can also be rewritten with explicit indication of the vector amplitude and phase:

$$\vec{V}[k] = M_V[k] \cdot e^{j\phi_V[k]} \tag{2.14}$$

The representation of signals in the complex plane will be used extensively throughout this work because of the compactness of its formulation. Before transmission of the signal vector as defined in (2.13) a conversion to the analog domain is necessary.

It has been shown that quadrature upconversion to an RF frequency can be realized fully on-chip using specific transmitter topologies that avoid the use of external components ([Crols PhD97, Rofoug JSSC98, Aero, Ajji ISSCC01]). Such a complete transmitter topology, starting from the Digital Signal Processor (DSP) that delivers the digitally modulated bit stream, is shown in Fig. 2.20. The digital I/Q pair is converted to an analog I/Q pair by a Digital-to-Analog (D/A) convertor and filtered by a low pass filter. After that, the analog I/Q signal is ready for upconversion. Similar to (2.13), a mathematical expression for the analog signal vector in the continuous (complex) time domain exists:

$$\vec{V}(t) = I(t) + j \cdot Q(t) \tag{2.15}$$

The goal of upconversion is clearly to transmit the signal $\vec{V}(t)$ at the carrier frequency f_c as shown in Fig. 2.18. Since both $I(t)$ and $Q(t)$ are real signals, the multiplication of $Q(t)$ by j must be implemented within the upconversion

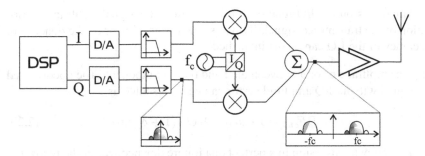

Figure 2.20. Transmitter Topology

operation. As can be seen in both topologies of Fig. 2.18 and Fig. 2.20, an
I/Q LO-signal is needed for quadrature upconversion. Here, the I/Q stands for
a phase shift of $\frac{\pi}{2}$ between the two LO signal components. And to complete
the story: a phase difference of $\frac{\pi}{2}$ in the complex domain is equivalent to a
multiplication with j.

Both a cosine and sine can be written in the complex domain as a sum of two
exponentials:

$$\cos \omega t = \frac{e^{j\omega t} + e^{-j\omega t}}{2} \qquad (2.16)$$

$$\cos\left(\omega t - \frac{\pi}{2}\right) = \sin \omega t = \frac{e^{j\omega t} - e^{-j\omega t}}{2j} \qquad (2.17)$$

From (2.17) it is clear that a cosine and a sine are "naturally" in quadrature,
since the sine is simply a cosine shifted over $\frac{\pi}{2}$.

Using the expressions derived above, the mathematical calculation can be
completed. A quadrature upconversion can be expressed as:

$$RF_{out} = I(t) \cdot \cos 2\pi f_c t + Q(t) \cdot \sin 2\pi f_c t \qquad (2.18)$$

Using (2.16), (2.17) and $\alpha_c = 2\pi f_c t$, following expression is derived:

$$RF_{out} = [I(t) + j \cdot Q(t)] \cdot \frac{e^{-j\alpha_c}}{2} + [I(t) - j \cdot Q(t)] \cdot \frac{e^{j\alpha_c}}{2} \qquad (2.19)$$

Thus, the wanted output is found at the negative carrier frequency. At the
positive carrier frequency the Q-component appears with a minus sign in the
complex sum in (2.19), corresponding to a phase shift of π. Due to this phase
shift, it is possible to recover the I and Q components from the transmitted
RF_{out} signal simply by multiplication with a cosine resp. sine in a quadrature
downconvertor. This downconversion path is beautifully dealt with in [Jans
PhD01], and will not be discussed here any further.

Any errors on the I/Q relationship in the oscillator signal result in deterioration of the transmitted signal. Two possible forms of error on the quadrature accuracy of the LO can be distinguished:

1 An amplitude error δ between the I and the Q component. The upconverted signal with an I/Q amplitude error can be represented as:

$$RF_{out} = I \cdot \cos \alpha_c + Q \cdot (1 + \delta) \sin \alpha_c \qquad (2.20)$$

After downconversion in a perfect quadrature downconvertor the recovered baseband signal becomes:

$$BB = I + j \cdot Q(1 + \delta) \qquad (2.21)$$

Thus, the amplitude error on the I/Q LO-signal in the transmitter is transferred to the baseband signal. In the frequency spectrum of the transmitted signal, this error manifests itself as "spectral regrowth", being a spectral mask that deviates a little from the theoretical mask of the used modulation type. This is shown in Fig. 2.21(a) by the spectrum in dashed line;

2 A phase error ϕ on the ideal phase difference of $\frac{\pi}{2}$ between the I and the Q component of the oscillator signal. The transmitted signal including phase error can be represented as:

$$I \cdot \cos \alpha_c + Q \cdot \sin (\alpha_c + \phi) \qquad (2.22)$$

The effect of it is that the multiplication in the complex domain of Q with j becomes a multiplication with je^ϕ. The recovered baseband signal after downconversion in a perfect quadrature downconvertor thus becomes:

$$BB = I + je^\phi \cdot Q \qquad (2.23)$$

As can be seen in Fig. 2.21(b), this error in fact introduces a rotation in the complex plane of the transmitted symbol vectors. In the frequency spectrum, this error will probably be invisible. However, it can seriously deteriorate the bit-error-rate (BER) of the system, especially in higher-order modulation schemes as QAM64 or QAM256 since there a small phase shift can already lead to the erroneous detection of a symbol that is a neighbor of the effectively transmitted symbol in the modulation mask.

Mostly, the allowable error is deduced from the system specifications of the applications for which the design is made. In the next two subsections, methods for quadrature LO generation are discussed, with an indication of the possible origin of I/Q errors.

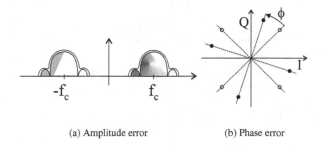

(a) Amplitude error   (b) Phase error

Figure 2.21.   Effects of quadrature errors in the LO

2.4.2   Polyphase Network

2.4.2.1   Harmonic Polyphase Signals

To be able to explain the properties of a polyphase *network*, first the definition of a polyphase *signal* is given. We will only consider harmonic polyphase
signals, consisting solely of harmonic signals. This does not undermine the
generality of what follows since using a Fourier decomposition, an arbitrary
waveform can be written as a sum of harmonic signals. To describe the signal components of a polyphase signal, we use the rotating vector, or *phasor*
representation of a sine wave. Let the start point of the rotating vector coincide with the origin of a complex coordinate system, as depicted in Fig. 2.22.
Then, the end-point of the vector with length A_p and start phase ϕ turning in an
anti-clockwise direction with constant angular velocity ω , moves in a *uniform
circular motion* that can be described mathematically as:

$$z(t) = A_p \cdot e^{j\omega t + \phi} \tag{2.24}$$

$$= A_p \cdot (\cos(\omega t + \phi) + j \cdot \sin(\omega t + \phi)) \tag{2.25}$$

$$= A_p e^{\phi} \cdot e^{j\omega t} = \bar{A} \cdot e^{j\omega t} \tag{2.26}$$

As (2.25) indicates, the projection of this vector on the real and imaginary axis
results in a cosine respectively sine wave, with amplitude A_p, angular frequency
ω and start phase ϕ. The physically meaningful part is of course the projection
on the real axis, thus reducing the representation in the complex plane to a cosine
wave. Since $\omega = 2\pi f$, with f the oscillation frequency, we can use either of
them to quantify the same property of the harmonic oscillation. Equation (2.26)
defines the complex vector $\bar{A}$, called *phasor*, that fully describes the harmonic
wave without explicit indication of the time and frequency dependent part. This
phasor representation is useful for calculations on a network in the sinusoidal
steady state, with only one sinusoidal frequency present.

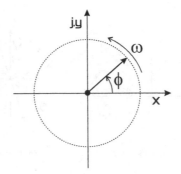

Figure 2.22.   Phasor representation of a harmonic oscillation

In general, a (harmonic) polyphase signal consists of M harmonic signals [9], with identical frequency but different start phase and possibly different amplitude. Because the signal components have the same frequency, the rotating vectors representing them have an identical angular velocity and thus the phasor representation can be used for calculations on the multi-component signal. A special subset of polyphase signals comprises those with equal amplitude and with start phases that are equally spaced over 360°. This subset is called the set of *symmetric* polyphase signals, since they form a symmetric constellation in phasor representation.

Some examples are given to show that a "polyphase" signal is in fact quite commonly used:

- A differential signal: it consists of a pair of two signal components ($M = 2$) with equal frequency and amplitude, but with a phase difference of 180°. This is a symmetric polyphase signal;

- A "perfect" quadrature signal: it has two signal components ($M = 2$) with equal base frequency and amplitude, that differ exactly 90° in phase. This is a *non*-symmetric polyphase signal, because the start phases are not equally spaced;

- A "perfect" differential quadrature signal: it has four signal components ($M = 4$) with equal base frequency, and a phase difference between two consecutive phasors of 90°. This is a symmetric polyphase signal.

Assuming the start phases of the consecutive phasors $1..M$ of an M-phase polyphase signal to be a monotone increasing or decreasing sequence, a dis-

---

[9]These sub-signals are also called *phases*, hence poly-phase. However, in this text we'll use the term *signal component* to indicate these sub-signals in order to avoid any confusion with the absolute or relative phase of the signal itself.

tinction can be made between a positive and a negative polarity of the polyphase signal. Obviously, the polarity of a 2-phase differential signal can not be specified. An example clarifies the two possible polarities in the case of a 4-phase signal.

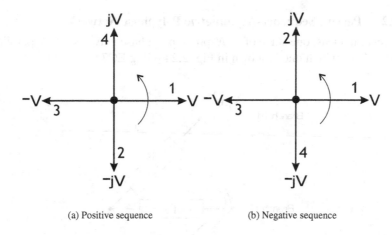

(a) Positive sequence   (b) Negative sequence

Figure 2.23. Positive and negative sequence of a 4-phase signal

Phase Sequence Example Fig. 2.23 shows the phasor representation of a positive sequence and of a negative sequence. Both are symmetric polyphase signals. The real axis is drawn in the horizontal direction. Fig. 2.23(a) is a positive sequence because each phasor with index i leads the phasor with index $i + 1$, in this case by 90°. Thus, the start phase has a monotone increasing value when going through the signal phasors 1..4. In other words: the start phase sequence (0°,90°,180°,270°) results in a polyphase signal with a positive polarity. The following phase sequence has a negative polarity, since each phasor with index i lags the phasor with index $i + 1$: (0°,-90°,(-)180°,-270°). Here, the start phase sequence is monotone decreasing.

Negative Frequencies A polyphase signal with a negative polarity, can also be regarded as having a negative *frequency*. Since a negative sequence can be distinguished by the fact that phasor i lags phasor $i + 1$ in phase, an inversion of the sense of rotation in Fig. 2.23(a), results in a polyphase signal with a negative sequence. Thus, changing the sense of rotation is equivalent to the change of order of the phasors as depicted in Fig. 2.23(b). Of course, we can never directly change the sense of rotation of the phasors, since this is physically impossible. However, the conclusion is that the use of a polyphase signal with a negative phase sequence to evaluate a circuit, is in fact the same as evaluating that circuit for negative frequencies.

Some circuits behave differently for positive or negative polarization of the polyphase signal. One of these circuits is a sequence asymmetric polyphase network, as will be discussed next.

2.4.2.2   Passive Sequence-Asymmetric Polyphase Network

In general terms, one stage of an N-phase polyphase network has N parallel branches[10], of which one is shown in Fig. 2.24 [Ging EC73].

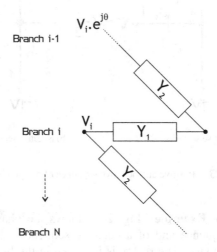

Figure 2.24.   One branch of a general polyphase network

Since the polyphase networks used in this work all function within a quadrature signalling system, a phase shift of 90° should exist between two adjacent branches. The phase shift between two adjacent branches of a N-phase polyphase network is, due to the perfect symmetry, equal to $\frac{360}{N}$ degrees, and therefore $N = 4$ branches are needed for a 90° phase difference. Fig. 2.25 shows a clear example of what is meant in the remainder of this text with a polyphase network.

In analogy with a single-phase network description, we want to express the network behavior of a polyphase network as one transfer function. However, a difference has to be made for positive and negative input sequences. For a single-phase network, one harmonic signal is used as input signal to calculate the output signal, and from that the transfer function. To derive the positive and negative

---

[10]For the terminology of a polyphase network, we'll use the term *branches* to indicate the different phases of the network. This again is done to avoid any confusion with the phase of the signal itself.

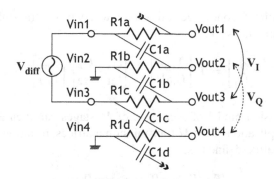

Figure 2.25. $N = 4$ polyphase network, for differential-to-quadrature conversion

transfer functions of a polyphase network, a symmetric, harmonic polyphase signal is applied to it. The response to non-symmetric, harmonic polyphase signals can be deduced from this transfer function by a vector-decomposition of the applied non-symmetric N_{ns}- signal into N_{ns} symmetric phasor signals [Ging EC73]. The response to non-harmonic polyphase signals can be derived by a Fourier decomposition of the input signal into harmonic polyphase signals, as is common practice for single-phase signals.

The branch elements of the polyphase network as used in this work are all passive, more specific a resistor is used to realize admittance Y1 and a capacitor is used to realize admittance Y2 of Fig. 2.24. This passive realization leads us to the explanation of the full name for such a circuit, being a "passive sequence-asymmetric polyphase network".

- Passive; because it consists of passive components only, i.c. resistors and capacitors;

- Sequence Asymmetric; this stems from the inherent property of a polyphase network that it has a transfer characteristic for a polyphase signal that depends on its polarity.

This asymmetric behavior will be dealt with in more detail in the next section.

2.4.2.3   Analytical Calculations on a Polyphase Network

For the single branch of a polyphase network as shown in Fig. 2.24 the chain matrix expressing the input/output behavior can be written down as [Ging EC73, Galal TCASII00]:

$$
\begin{bmatrix} V_{in} \\ I_{in} \end{bmatrix} = \frac{1}{Y_1 + e^{j\theta}Y_2} \begin{bmatrix} Y_1 + Y_2 & 1 \\ 2Y_1Y_2(1 - \cos\theta) & Y_1 + Y_2 \end{bmatrix} \begin{bmatrix} V_{out} \\ I_{out} \end{bmatrix} \tag{2.27}
$$

This formulation for one stage corresponds to a two-port transmission matrix representation of a network:

$$\begin{bmatrix} V_{in} \\ I_{in} \end{bmatrix} = \mathbf{T} \begin{bmatrix} V_{out} \\ I_{out} \end{bmatrix}^{11} = \begin{bmatrix} t_{11} & t_{12} \\ t_{21} & t_{22} \end{bmatrix} \begin{bmatrix} V_{out} \\ I_{out} \end{bmatrix} \qquad (2.28)$$

The transmission model of a cascade of M stages can then easily be obtained by multiplication of the M transmission matrices, resulting in an overall transmission matrix, defined as:

$$\mathbf{T} = \mathbf{T}_{(1)} * \mathbf{T}_{(2)} * \cdots * \mathbf{T}_{(M)} \qquad (2.29)$$

From this transmission representation of the multi-stage polyphase network, the transfer function can be calculated for two distinct cases:

- $\mathbf{I_{out}} = \mathbf{0}$ This assumption holds for an open circuit at the output of the polyphase network. The resulting transfer function thus describes the *voltage* input/output behavior of the polyphase network in the frequency domain;

- $\mathbf{V_{out}} = \mathbf{0}$ This assumption holds for a short circuit at the output of the polyphase network. The resulting transfer function thus describes the *current* input/output behavior of the polyphase network in the frequency domain.

For other cases, no simple v_{out}/v_{in} or i_{out}/i_{in} relation can be formulated without taking input and output circuitry into account. The influence of loading effects on the behavior of a polyphase network is elaborated in Section 2.4.2.5. It will be shown that the quadrature image ratio is not influenced by the output loading.

To calculate the voltage ratio of the quadrature-phase output (V_Q) to the in-phase output (V_I) for a polyphase network connected to a differential signal as in Fig. 2.25, following procedure is used:

1. The admittance symbols Y_1 and Y_2 are assigned as: $Y_1 = 1/R$ and $Y_2 = sC$ to reflect the used RC-topology of Fig. 2.25;

2. As stated above, for a polyphase network with multiple stages the total chain matrix can be calculated by simple matrix multiplications of the chain matrices of the consecutive stages;

---

[11] In a standard two-port description, a negative sign is used for I_{out} [Chua 87]. The convention used in this work is that the current I_{in} flows into the circuit and I_{out} flows out. For I_{out} this means the current flows in an inverse direction when compared to the 'normal' two-port description.

3 The matrices for positive and negative input sequences can be deduced from (2.27), by replacing θ with the effective phase-difference between two branches. Since we use only polyphase networks with 90° phase differences, θ will have an absolute value of $\frac{\pi}{2}$. For a positive input sequence, the signal contribution of branch $i-1$ in Fig. 2.24 through $Y2$ on the output of branch i will *lead* the signal contribution of branch i through $Y1$ in phase. Therefore, a positive input sequence will have a $\theta = -\frac{\pi}{2}$ and consequently a negative input sequence will have a $\theta = +\frac{\pi}{2}$;

4 Two different chain matrices are obtained: one for positive input sequences and one for negative input sequences;

   Example To clarify this, the chain matrix for a positive input sequence $(\theta = -\frac{\pi}{2})$ is given here:

$$\begin{bmatrix} V_{in} \\ I_{in} \end{bmatrix} = \frac{1}{1 - jsRC} \begin{bmatrix} 1 + sRC & R \\ 2sC & 1 + sRC \end{bmatrix} \begin{bmatrix} V_{out} \\ I_{out} \end{bmatrix}$$

5 For the calculation of the two $H = V_{out}/V_{in}$ voltage transfer characteristics, a zero output current I_{out} (i.e. open circuit) is assumed;

   Example The transfer function $H(\omega)$ for positive input sequences, is given by:

$$H(\omega) = \frac{V_{out}}{V_{in}} = \frac{1 + \omega RC}{1 + sRC}$$

   and for negative input sequences:

$$H(-\omega) = \frac{V_{out}}{V_{in}} = \frac{1 - \omega RC}{1 + sRC}$$

6 The two resulting transfer characteristics are viable for a single branch of the network and for a single component of a four-phase signal with either a positive or a negative input sequence. To calculate the output for a differential (two-phase) signal, the signal is represented as the sum of two four-phase signals, with opposite sequences. This is done using a vector decomposition as represented in Fig. 2.26, in which the real axis is pointing up and the imaginary axis is pointing to the left;

   Example The value of V_{in2} of Fig. 2.25 is thus represented as

$$V_{in2} = -j\frac{V}{2} \cdot e^{j\omega t} + j\frac{V}{2} \cdot e^{-j\omega t}$$

7 Since the polyphase network is a linear system, the superposition principle holds. Thus, the output of each branch can be calculated as the sum of

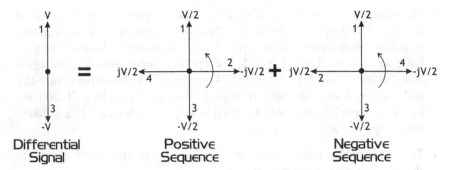

Differential Signal Positive Sequence Negative Sequence

Figure 2.26.   Phasor decomposition of a differential signal

the individual outputs generated by a positive sequence on the one hand and from a negative sequence on the other. Both have a different transfer characteristic, derived in step 5;

Example   The result for output V_{out2} of Fig. 2.25 is calculated as:

$$V_{out2} = -j\frac{V}{2}(H(\omega) \cdot e^{j\omega t} - H(-\omega) \cdot e^{-j\omega t})$$

8 The outputs V_I and V_Q in Fig. 2.25 can then be calculated as:
$V_I = V_{out1} - V_{out3}$ and $V_Q = V_{out2} - V_{out4}$

Following this flow, the analytical expression for the relation between the I and Q components at the output of a polyphase network with three stages, driven by a differential input is derived [Galal TCASII00]:

$$\frac{V_Q(s)}{V_I(s)} = \frac{s(R_1C_1 + R_2C_2 + R_3C_3) - s^3(R_1R_2R_3C_1C_2C_3)}{1 - s^2(R_1R_2C_1C_2 + R_1R_3C_1C_3 + R_2R_3C_2C_3)} \qquad (2.30)$$

For four stages the calculated formula is:

$$\frac{V_Q(s)}{V_I(s)} = \frac{as - bs^3}{1 - cs^2 + ds^4} \qquad (2.31)$$

with

$$a = R_1C_1 + R_2C_2 + R_3C_3 + R_4C_4$$
$$b = R_1R_2R_3C_1C_2C_3 + R_2R_3R_4C_2C_3C_4$$
$$+ R_1R_2R_4C_1C_2C_4 + R_1R_3R_4C_1C_3C_4$$
$$c = R_1R_2C_1C_2 + R_1R_3C_1C_3 + R_1R_4C_1C_4$$
$$+ R_2R_3C_2C_3 + R_2R_4C_2C_4 + R_3R_4C_3C_4$$
$$d = R_1R_2R_3R_4C_1C_2C_3C_4$$

The phase and magnitude of formulae (2.30) and (2.31) can be interpreted as follows:

- The phase indicates the phase difference between the I and Q outputs as a function of frequency. The deviation of this difference from the ideal 90° is called the *Phase Error* ϕ;

- The magnitude Δ is a measure for the amplitude difference between the I and Q outputs as a function of frequency. The deviation from zero is called the *Gain Error* $= \Delta - 1$.

2.4.2.4   Image Ratio

Both Phase and Gain Error have an influence on the final Image Ratio (IR) that can be obtained when the quadrature signal is used in transceiver topologies as described above in Section 2.4.1. To calculate the Image Ratio after the mixing process within a generic transceiver topology, an analytical formula exists [Arch JSSC81]:

$$IR = \frac{A^2 - 2AB\cos(\phi + \theta) + B^2}{A^2 + 2AB\cos(\phi + \theta) + B^2} \tag{2.32}$$

with A and B the amplitudes of the voltage gains of the I and Q channel respectively, ϕ the Phase Error of the LO-signal and θ the Phase Error on the information signal. This formula can be simplified to reflect only the effect of the Gain and Phase Errors of the polyphase network, thus considering the other components to be flawless. This leads to the following formula for the Image Ratio at the output of the polyphase network:

$$IR = \frac{1 - 2\Delta\cos\phi + \Delta^2}{1 + 2\Delta\cos\phi + \Delta^2} \tag{2.33}$$

with Δ the magnitude of the ratio between I and Q channel, and ϕ the Phase Error between the I and the Q channel of the polyphase network. As mentioned above these are calculated from (2.30) or (2.31). For a four stage polyphase network with values for resistors and capacitors as depicted in Table 2.4, a plot of the obtainable IR without mismatch is given as a function of frequency in Fig. 2.27.

Mismatch is not taken into account, because this can not be dealt with using the analytical approach presented above. Therefore, the simulation techniques for a polyphase network that is used as a quadrature generator will be explained in a following section. First the effects of loading and the calculation of the input impedance of a polyphase network will be discussed.

Table 2.4.   R&C Values of the four stage polyphase network used for simulations

resistor	value [Ω]	capacitor	value [fF]
R_1	295	C_1	800
R_2	200	C_2	800
R_3	110	C_3	800
R_4	75	C_4	800

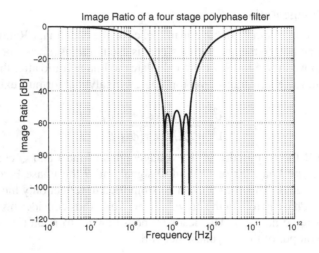

Figure 2.27.   IR for a four stage polyphase network (no mismatch)

2.4.2.5   Loading Effects & Input Impedance

The image ratio as calculated above, is insensitive to the effects of output loading, when the loads are identical for all four outputs. As explained in more detail in Section 4.1.2.3, the influences of loading can be calculated using a loaded two port representation of the polyphase network. Since the IR is calculated based on a difference between network outputs, an identical loading of all outputs will not result in any difference for the IR.

However, the *input impedance* does experience influence from the output load. In Fig. 2.28 the input impedance of a four stage polyphase network with component values given in Table 2.4, loaded with a resistor *Rload* or capacitor *Cload*, is shown. The plots have been obtained using the detailed calculations that are derived in Section 4.1.2.3. It is clear that the plot of the input impedance versus frequency has a rather irregular shape. This has to be taken into account in the design of the input buffer.

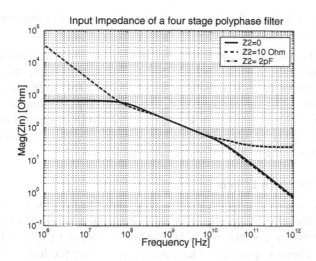

Figure 2.28. Zin of a loaded four stage polyphase network

For the differentially driven polyphase network used as quadrature generator (Fig. 2.25) an additional remark has to be made. The AC ground connection of the two unused input terminals influences the input impedance. This is caused by the fact that the R/C splitter block that is present at each of these input terminals has a non-symmetrical behavior at frequencies different from the RC-pole frequency. For such frequencies, the impedance as seen into the resistive part will always differ from the impedance as seen into the capacitive part. The current flowing from the positive input terminal Vin1 into capacitor C1a *(thus flowing "down" looking at Fig. 2.25)* and through resistor R1b to node Vin2 will differ from the current flowing from the negative input terminal Vin3 through the resistive branch R1c and capacitor C1b with an opposite direction into that same node Vin2. Thus, a net current will flow out of terminal Vin2 into the AC ground, causing the input impedance seen at node Vin1 to be *lower* than for a polyphase network with all four input terminals connected to an AC signal. A similar argumentation holds for input terminal Vin3. This reasoning also shows why an AC connection to ground at input terminals Vin2 and Vin4 is necessary to ensure a proper operation of the polyphase network as a differential-to-quadrature convertor. Omitting this connection would lead to an unbalance in the polyphase network behavior, resulting in inaccurate quadrature generation.

2.4.2.6   Simulation of a Polyphase Network used as Quadrature Generator

Using a simulation scheme as depicted in Fig. 2.25, with a differential input voltage source at input terminals 1 and 3 of a polyphase network, and taking the differential output voltage at output terminals 1 and 3 for the in-phase component V_I, and the differential output voltage between output terminals 2 and 4 for the quadrature-phase component V_Q, the image ratio IR as defined above, can be calculated using (2.32), with $A = \text{magnitude}(V_I)$, $B = \text{magnitude}(V_Q)$, $\theta = 0$ and $\phi + \frac{\pi}{2}$ the phase difference between V_I and V_Q.

2.4.2.7   Mismatch Effects

The operation of a polyphase network as described up till now, does not incorporate the effect of component mismatch. In a fabricated design, the mismatch between components is impossible to rule out completely, unless some kind of trimming or tuning technique is used after production. Since the cost of this extra production step is often unacceptable due to the low profit margins on the final products, we do not take it under consideration.

Therefore, the effects of mismatch on the performance of the polyphase network have to be taken into account from the beginning of the design. In the analytical approach described in the previous paragraph no matching information can be included directly. To investigate the effects of mismatch on circuit performance, generally the Monte Carlo simulation method is used ([Papo 84]). A batch of simulations is performed, selecting for each single simulation a set of random values for all mismatch-sensitive components. The random value for a single component is chosen out of a Gaussian distribution, with a standard mean equal to the nominal designed value of the component and a standard deviation that is calculated based on the mismatch data made available by the technology provider. The resulting set of values for each output variable *(voltage, current, noise voltage, ...)* or related performance variable *(noise figure, HD2, IIP3,...)* mostly also shows a Gaussian distribution of which the standard mean and standard deviation can be derived. From these distributions, the yield with respect to a certain performance variable can be calculated as the area underneath the Gaussian distribution corresponding with acceptable values of that specific performance variable.

> **Example** The yield for an opamp circuit with a desired gain of 60dB ±0.5dB is simply the area of the gain-value distribution for which the gain is between 59.5dB and 60.5dB.

This Monte Carlo technique is a valuable approach for circuits with many elements of which the individual contribution of mismatch on the behavior is not clear. However, since this technique is based on random sequences its reliability (or credibility) for the prediction of circuit behavior, is dependent on the implementation of the technique and the "quality" of the used random sequence(s) [Pawl Comm02]. Moreover, "... the most prudent policy for a person to follow is to run each Monte Carlo program at least twice, using quite different sources of pseudo-random numbers, before taking the answers of the program seriously" [Knuth 98].

A specific drawback for a polyphase network is that the needed number of evaluations to obtain a reliable result is relatively high, compared with the simplicity of the network. Considering the high symmetry of a polyphase network, two other solutions to obtain a good estimation of the filter behavior with inclusion of mismatch, exist:

1 An approximate analytical calculation of the image frequency output due to the resistor and capacitor mismatch;

2 Worst-case simulations of the polyphase network starting from a simulation-supported manual sensitivity analysis, incorporating the component mismatch directly in the circuit network.

The first solution is calculated in [Beh JSSC01] for one stage of a polyphase network and is based on the following assumptions:

- The four signal paths are uncorrelated;

- The input frequency is not exactly equal to the pole frequency (at that exact frequency, the mismatch has no influence);

- The distribution of the resistor value is Gaussian, with mean value R and standard deviation σ_R;

- The distribution of the capacitor value is Gaussian, with mean value C and standard deviation σ_C;

- The final image due to mismatch has a Gaussian distribution with zero mean and standard deviation $\sigma(\text{Image Out})$.

The resulting formula directly relates the maximum attainable image ratio of a one stage polyphase network to the component mismatch:

$$\frac{\sigma(\text{Image Out})}{\text{Desired Out}} = \frac{1}{4}\sqrt{\left(\frac{\sigma_R}{R}\right)^2 + \left(\frac{\sigma_C}{C}\right)^2} \qquad (2.34)$$

Example Suppose we want to design a polyphase network with a guaranteed image suppression of 50dB or better, and this with a yield of $3.\sigma$. This means that 99.9% of the produced circuits will have an image suppression that is better than 50dB, corresponding to the total area underneath a normalized Gaussian distribution from $-\infty$ to 3. The total sigma of the filter thus has to be lower than $-50dB/3 = 1.054e - 3$. Assuming that the sigma of the capacitors and the resistors in (2.34) is equal, this leads to a value of $\frac{\sigma_R}{R} = \frac{\sigma_C}{C} \leq 0.298\%$.

The second approach uses a work flow that enables us to choose the expected yield, expressed as a number of times the standard deviation of the analog block under consideration and then simulate this analog block to obtain the worst-case behavior we can expect within the specified yield. E.g. 99% (=Yield) of the fabricated circuits should have a behavior that is better than or equal to the simulated behavior. Following assumptions are made to come to the final work flow:

1 Consecutive stages have an uncorrelated influence on the total mismatch. This means that the standard deviation of the complete n-stage polyphase network can be expressed as: $\sigma_{tot} = \sqrt{n} \cdot \sigma_{single}$, assuming the standard deviation of all stages are equal. Since this is a statistical consideration and calculation, we have to make sure that we include it in the simulations. This is done by using $\frac{\sigma_{single}}{\sqrt{n}}$ as the worst-case deviation of a single stage with a standard deviation σ_{single}, within an n-stage polyphase network. This can be interpreted as if the parameter value distribution is "narrowed";

2 The worst case for the total mismatch of the polyphase network is obtained when all stages of the polyphase network have a worst-case contribution. Therefore, the alteration made to each stage to reflect the worst-case situation can be similar for each consecutive stage;

3 In the sensitivity analysis, the worst case for one phase in a single stage is that both resistor and capacitor have a deviation of their value that act in a similar way on the pole frequency they define together. E.g. if the resistor is bigger than the nominal value, the worst effect on the pole frequency will be seen for a capacitor that is also bigger than the nominal value.

Combination of the first two assumptions, leads to the constatation that a sensitivity analysis on mismatch of a single-stage polyphase network is sufficient to build-up a mismatch-simulation circuit for a polyphase network with an indefinite number of stages. To facilitate this build-up, a parameterized base cell for one stage of a polyphase network is used. The sensitivity analysis itself is done employing manual circuit adaptations to this base cell and using a circuit simulator. Define σ_R and σ_C as the standard deviation of the component value of respectively resistor and capacitor in the one stage polyphase network. Then the adaptations used in the sensitivity analysis consist of the addition (or subtraction) of a number of times σ_R or σ_C to the nominal component value. Assumption three already restricts the number of combinations that have to be tried out. The result of the sensitivity analysis is the adapted circuit of Fig. 2.29 that incorporates the worst possible combination of parameters.

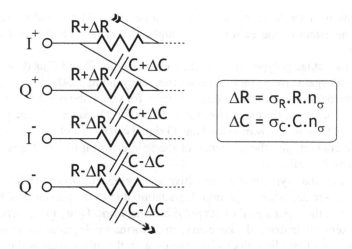

$$\Delta R = \sigma_R.R.n_\sigma$$
$$\Delta C = \sigma_C.C.n_\sigma$$

Figure 2.29. Circuit for simulation of a polyphase network including mismatch

Using this adapted circuit, the four stage polyphase network of a CMOS upconvertor has been designed. It is presented in Chapter 5 and a plot of the simulated Image Ratio that can be obtained with the generated quadrature LO is given in Fig. 5.10.

2.4.2.8 Quadrature Unbalance for Higher Harmonics

Although a polyphase filter can be designed to supply a quadrature signal for a broad range of input frequencies, mostly it is designed for the frequency band of the *first harmonic* of the local oscillator signal. Consequently, any higher harmonics fall out of the quadrature generating band of the polyphase network. These higher harmonics will have a non-quadrature relation at the output of the network. For the mixing block following the polyphase filter, this can pose severe limitations on the obtainable image rejection.

At the output of the polyphase network, the phase difference is a constant 90 degrees above a certain frequency when no mismatch is present. Since the effect of mismatch on the phase difference is quite constant over a broad frequency range ([Galal TCASII00]), the phase difference between I and Q output of the polyphase network does not differ substantially for higher harmonics compared to the first harmonic, even with mismatch present. Therefore, the effect of component mismatch on the quadrature *phase* accuracy can be taken into account completely by simply looking at the influence of mismatch on the phase accuracy for the first harmonic.

The *amplitude* difference between I and Q, resulting from the gain error, could be removed by clipping both outputs. However, this solution may introduce additional phase unbalance, and will certainly lead to a higher power consumption. Therefore, this solution is not considered here, and we have to look at the effect of the gain error for higher harmonics on the final image rejection.

For a four stage polyphase network, with values for R and C as depicted in Table 2.4, the gain error is plotted versus frequency in Fig. 2.30. In Fig. 2.30(a), the gain error is shown for the frequency band for which the polyphase filter is designed whereas in (b) the gain error is shown for a broader frequency band. Using (2.31), the total power in the I and Q signal can be calculated for a certain harmonic content, and the influence of the gain error on the final quadrature balance can be evaluated.

This way, mixer types that are sensitive to the global quadrature balance can suffer from serious image rejection degradation due to the presence of higher harmonics in the input signal of the polyphase network. Other mixer types like linear mixers will in first order have no such performance degradation. However, non-idealities of building blocks interfacing with the mixer such as the output buffer of a linear mixer topology, can deteriorate the mixer performance due to second-order effects as will be discussed in Section 5.3.5.

2.4.2.9   Design Considerations

In the physical design of a polyphase network, some practical considerations have to be taken into account. A non-limiting overview is given here.

Parasitic Capacitance   Both the resistors and the capacitors in the polyphase network have a certain parasitic capacitance towards the substrate. The substrate itself is connected to ground by substrate straps. Both the substrate resistivity and the physical distances from the parasitic capacitor to the nearest substrate straps define the series resistance of the connection through the substrate between the parasitic capacitor and the signal ground on chip. The exact calculation of the effective series resistance is not straightforward because, in general, the matrix of substrate straps forms a complex resistive network.

The series-connected capacitor and resistor form a complex impedance that loads the polyphase network. Due to the series resistor, the voltage drop over the capacitor will be lower, and thus the effective capacitance as seen by the polyphase network is smaller than in the absence of the series resistor. To calculate the effective capacitance at a certain frequency, the series network is converted into a parallel network valid around that frequency consisting of a parallel resistor and capacitor. This representation enables us to make a distinction between the pure resistive loss into the parallel substrate resistance on the one hand and the phase shift and capacitive coupling caused by the capacitive component of the parasitic load on the other hand.

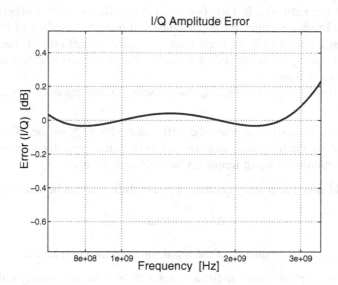

(a) In the useful frequency band

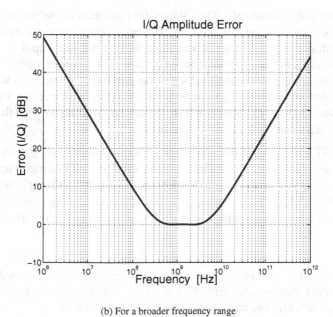

(b) For a broader frequency range

Figure 2.30. Gain error of a polyphase network

The capacitive coupling itself has an influence on the substrate noise sensitivity of the network. In fact, two conflicting effects exist: a bigger series resistance lowers the effective capacitance and thus the coupling of the noise voltage into the network but it generates a bigger (white) noise voltage from itself. An optimal value exists for the series resistance to come to a minimal substrate noise injection[12].

Apart from the noise injection, the presence of the parasitic capacitances results in:

- Additional signal loss, when the 3dB bandwidth f_{3dB}^{Par} of the RC-network formed by the polyphase resistor and its parasitic capacitor is comparable to the maximum signal frequency sent through the network;

- An additional frequency dependent phase shift, that can be calculated as

$$\Delta(\phi) = \arctan \frac{f_{input}}{f_{3dB}^{Par}}$$

  with f_{input} the frequency of the input signal of the polyphase network;

- Additional phase and amplitude mismatch, due to the (mostly unknown) mismatch on this parasitic capacitance.

Current or voltage output The input/output quantity of the polyphase network can be either a voltage or a current. The calculations above have been done in the assumption of a voltage driven input and an infinite output impedance drawing no current from the polyphase network. For a current output, the calculations are similar as in Section 2.4.2.3, but instead of setting I_{out} to zero in step 5, the output voltage V_{out} must be set to zero (i.e. short circuit). In CMOS design, two practical situations can be discerned, depending on what the output of the polyphase network is connected to:

- A MOST gate. This is the case if the network is followed by a source follower, an amplifier stage or directly connected to a MOST mixer transistor. Since no current flows into the MOST gate, an open circuit approach can be used. For high-frequency operation, the capacitive loading effect can not be neglected. In circuit simulations, this loading effect is automatically taken into account;

- A MOST source. Normally, this represents a low impedance for the polyphase network. Thus, the zero output voltage approximation can be used for the analytical calculations. Here, the non-zero impedance of the buffer circuit causes a resistive degeneration of the polyphase network.

---

[12] Of course, this is not the coupling of substrate noise coming from digital circuitry on the same die, which would be roughly proportional to the parallel capacitance value.

Varying Gain Error over suppress band   The Gain Error is not necessarily the same for a frequency $f_{cent} + \Delta(f)$ as for a frequency $f_{cent} - \Delta(f)$, with f_{cent} the center of the image suppress band of the polyphase network. Therefore, the image ratio will differ over the suppress band.

2.4.2.10   Final Design Flow

After all these considerations, a final design flow can be deduced. The design of a polyphase network will always be an iterative process. Since a thorough study of its behavior is possible with the methods described above, a Design Template as defined further on in Chapter 3 could be derived to come to an automated design optimization of the polyphase network and its interfacing functional blocks. Alternatively, the optimization of the polyphase networks can be done in Matlab ([Matlab]) using manually derived models. Irrespectively of the used optimization methodology, following issues have to be taken into account:

- The minimal capacitor size and value used in the polyphase network is defined by mismatch;

- The maximal resistor value that can be used is defined by its noise contribution;

- The maximal resistor size is restricted by the minimum 3 dB frequency of the RC network that the resistor forms with its parasitic capacitance;

- If I_{out}="0" *(voltage output)*, a voltage division occurs between the output impedance of the input buffer, the network impedance and the (ideally infinite) input impedance of the output buffer. Thus, the signal loss over the network can be estimated as function of these three variables. A higher network impedance or output impedance of the input buffer results in a higher voltage loss. In first order, the network impedance is proportional to its total resistance;

- If V_{out}= "0" *(current output)*, a higher network impedance will necessitate a higher voltage amplitude at the input to keep the (AC) current output constant. This will lead to more stringent requirements for the matching and linearity specifications of the input buffers;

- For either voltage output or current output, the output impedance of the input buffer must be kept low compared to the total network impedance to reduce loss. Thus, a network impedance that is too low will result in extreme low impedance demands for the input buffer, leading to high power consumption.

From the above it is clear that both input and output buffer have to be designed together with the polyphase network. Moreover, some specifications of the two buffers also depend on the circuits they are interfacing with. This clearly demonstrates the difficulty of good partitioning in analog design. It also underlines the need for design automation to deal with a set of entangled *Functional Blocks*[13] for which the functional description of each block and of the whole can be clearly formulated, but for which the circuit realization in terms of optimal circuit design is much more of a trial-and-error time occupancy than an analytically tractable and solvable problem that could be regarded as a real "design challenge" for an analog designer.

2.4.3   Frequency Dividers

Besides a polyphase network, a commonly used method to generate an accurate quadrature LO-signal, is based on the use of a divider circuit.

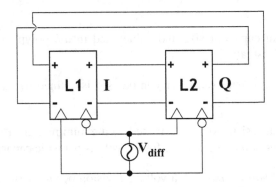

Figure 2.31.   Quadrature generation using a D-Flip-flop

Fig. 2.31 shows the block schematic of a digital divider used as a quadrature generator. A differential oscillator signal coming for instance from a VCO, is applied to the clock-input of a differential D-flip-flop, consisting of two edge-triggered latches (L1 and L2). L1 is connected to the positive clock signal (positive-edge triggered), L2 is connected to the inverted clock signal (negative-edge triggered). Since the clock signal is differential, this is done quite elegantly by simply inverting the clock connections as compared to L1. The output of L1 is used as input of L2 maintaining the signal polarity whereas the output of L2 is fed back to the input of L1 with *inversion* of the signal polarity. Thus, a phase shift of 180° is obtained in the loop consisting of the two latches. Since the loop gain is certainly higher than 1, the circuit is inherently unstable. A signal diagram of the D-flip-flop in operation is depicted in Fig. 2.32.

---

[13] as defined in Chapter 3

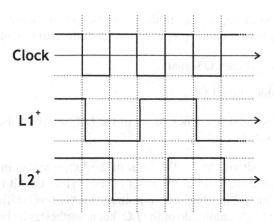

Figure 2.32. Signal waveforms of Fig. 2.31

Since the latches are made identical, and assuming also identical loads for both of them, their delay times are also the same. The result is that both the input clock frequency and the 180° phase difference over the divider loop, are divided exactly by two. Thus, a quadrature signal at half the input frequency of the clock is generated.

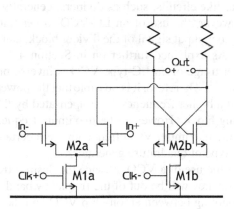

Figure 2.33. CMOS D-flip-flop implementation (one latch shown)

In Fig. 2.33 a possible CMOS circuit implementation of one latch of a D-flip-flop is given. It consists of two cross-coupled differential pairs [DeMuer ESSCIRC00]. The transistor sizing of this circuit is based on simulation, since no usable analytical approach can be found to describe its large-signal behavior. The positive and negative clock input terminals $Clk+$ and $Clk-$ are simply connected to the differential outputs of the LO. Three properties of this differential signal are of importance in the sizing step of the divider circuit, since they have a big impact on the proper operation of the building block :

- The DC level of the LO signal;

- The amplitude of the LO;

- The frequency and frequency range of the LO for which the flip-flop and the divider should work.

Due to the sensitivity to the DC level of the LO, often an implementation with an adjustable DC level at the input of the divider circuit is chosen. An example of such a DC-level adjuster is a simple MOST level shifter. The circuit is less sensitive to the amplitude of the LO, but nevertheless it should be within a certain range of acceptable values. The frequency has a big impact on the sizes that have to be chosen for the transistors of the circuit. Mostly, this type of divider circuit will only be operational within a certain frequency range. Therefore, a redesign is needed for every new frequency band the building block is to be used in.

The use of a frequency divider approach for quadrature generation calls for an input frequency of twice the wanted frequency of the final quadrature signal. This results in a power penalty for the divider block itself, since a higher frequency in digital-like circuits, such as dividers, generally leads to a higher current drain. However, the use of an LC-VCO to generate the differential signal used as the clock input signal of the divider block, can reduce the power penalty involved. As explained further on in Section 4.2, one can say that generally the power usage of an LC-type VCO is inverse proportional to the oscillation frequency[14]. Therefore, it is possible that the power increase to drive the divider block at a higher frequency is compensated by the power decrease in the VCO building block. However, the maximum frequency a divider can work at will limit the maximum VCO frequency for a certain technology that can be used in this type of quadrature generators.

An advantage of the use of a VCO at twice the desired frequency is that in most cases this frequency will be out of the frequency band of interest. Then, the unavoidable coupling between an on-chip VCO and the rest of the circuit will be less of a problem. However, when the frequency band of interest is so large that also the double of the needed LO frequency still lies within the band, this advantage no longer stands.

---

[14]For very high frequencies, this rule is not generally applicable anymore, as will be explained thoroughly in Section 4.2.3

One of the problems in the use of a divider as quadrature generator is a good estimate of the quadrature accuracy that can be obtained. If the duty cycle of the differential clock signal is exactly 50%, the major source of quadrature inaccuracy is the delay mismatch between the two latches within the divider block. Similarly as for the polyphase network, an adapted circuit with a worst-case performance under a certain amount of mismatch can be derived for the divider circuit. Also here, this can be done using a manual sensitivity analysis. However, the number of elements in this circuit of Fig. 2.33 is already larger than for the one stage polyphase network. Therefore, this method is a little more cumbersome than for the polyphase network. Mismatch here results in:

- Clock feedthrough, due to mismatch between the clock transistors M1;

- Quadrature inaccuracy, due to mismatch between the resistors and between the differential pairs M2.

A possible simulation approach is to construct an ideal quadrature upconversion mixer using non-linear voltage-controlled current sources that share a common load resistance. The I-output voltage of the divider is multiplied in such a source with an ideal (baseband) cosine and the Q-output voltage is multiplied with an ideal (baseband) sine. The Fourier spectrum of the voltage signal on the load resistance than has to be inspected for unwanted frequency components.

If the duty cycle of the differential clock signal differs from the ideal 50%, this directly results in I/Q inaccuracy. Mostly, the duty cycle of the VCO itself will be close to 50%, since the wave form of the VCO should be symmetrical to lower noise upconversion ([Haji JSSC98]). However, an LO buffer is inserted between the VCO and the divider to lower the capacitive load on the LC tank and to minimize kick-back effects from the switching divider on the LC tank, which would deteriorate the phase noise. A badly designed buffer can introduce a lot of second order harmonics. In other words: it can produce a wave form with a duty cycle far from the ideal 50%. Therefore, the design of the LO buffer between VCO and divider needs special attention if high I/Q accuracy is required.

2.4.4 Conclusion

Two alternative approaches for the generation of quadrature signals have been discussed in this section. The first is a differential oscillator followed by a polyphase network, the second a diver-type quadrature generator driven by a VCO oscillating at twice the wanted signal frequency. The choice depends on the available technology with its specific maximum operation frequency and mismatch characteristics for transistors and passive components and the total power consumption. In the calculation of the power consumption, all necessary buffers to obtain the wanted DC-level and amplitude have to be included. Only then a good and fair comparison between the two options can be made.

An automated design flow of the used building blocks with an optimization on system level, can help to evaluate both approaches within a reasonable design time. The focus of the designer then shifts from transistor level to system level, without losing any detail of the designed circuit. Moreover, also the layout level has to be included in the design flow since for high-frequency applications, the influence of parasitic elements in the layout can be detrimental for the system performance. This even reinforces the need for automation in the design flow, because a double design including layout is very unlikely to be considered in a professional environment where design time is one of the most valuable resources within a complete project. In Chapter 3 this need for design automation will be dealt with in detail. For now, it suffices to say that the need for a general design framework capable of system-level optimization introduced itself based on the very simple case one has to choose between two topologies for on-chip quadrature generation.

2.5   Manufacturing

The aspects of manufacturing this section deals with, is the part after the chip has been processed and the wafers cut into bare die. A fully-functional printed circuit board (PCB) with an electrically and mechanically connected chip is regarded here as the final product of the manufacturing step. The only electrical contacts that are available to connect to the circuits on chip are the bondpads that are foreseen during the design and layout of the chip. To be able to use the chip in a final product or perform tests on it, it is attached to some kind of carrier. This can be a package, a (test) substrate, another chip or the PCB itself. Independent of the kind of chip carrier, the electrical contacts on chip have to be connected to the conducting leads on the carrier. When the chip carrier is not the PCB, these conducting leads are on there turn connected to the printed lines on the PCB.

This section deals with one specific kind of *bonding process* that connects the bondpads with the conducting leads on the external chip carrier. Since different techniques exist to perform the bonding, a brief overview is given in a first subsection. The second section then deals with the flip-chip bonding technique as it has been used for the upconvertor chip described in Chapter 5.

2.5.1   The Bonding Process

The goal of the bonding process is to construct an electrically conducting connection between the bondpads on chip and the conducting leads on the chip carrier. In this text, two techniques are used:

■ Bondwire bonding. This is the most commonly used technique. The connection is made by attaching a gold or aluminum wire using (thermo)compression
   to the bondpads and to the electrical leads. When gold is used, a ball-wedge

technique is mostly used and for aluminum bondwires, a wedge-wedge technique is used. A sample of the VCO presented in Section 4.2 that has been wire-bonded to an alumina substrate is shown in Fig. 2.34;

- Flip-chip bonding. The electrical connection is made here by electrically conducting "bumps". In a first phase, these bumps are put on the bondpads on chip or on leads of the chip carrier. Then, the chip is turned over (hence: flip-chip) and an electrical and mechanical connection is made between chip and chip carrier using a "joining" process that is dependent of the used type of bump. Main advantages of flip-chip are the reduced bondwire parasitics [Carc PhD01] and the fact that all bonds are made in one joining step. Of course, the latter results only in a real gain in manufacturing time if also the bumps are applied in one step.

Figure 2.34. Wire-bonded chip

In general, the minimal pitch being the sum of the minimal distance and the size of a conductor, is smaller on chip than on the chip carrier, and much smaller than on the PCB. Thus, normally the connection area on the chip carrier will be larger than the area used by the bondpads on chip. However, when a flip-chip technique is used, also on the carrier bondpads with the same pitch as on chip have to be foreseen. This can be realized by using a small-pitch manufacturing process for the connection area on the carrier and by taking into account the minimal pitch of this manufacturing process for the placement of the bondpads on chip. Thus, the bondpad distance for a flip-chip bonded chip will be larger than the minimal distance as set by the electrical design rules of the chip technology.

For the flip-chip process, different types of bumps exist with their specific joining process to connect the bumped chip to the leads on the chip carrier:

■ Solder bumps. Different techniques exist to apply this type of bumps. They require a "wettable" surface layer on top of the normal Au or Al bondpads on chip, in order to allow the solder to make contact with the bondpad. A reflow solder process in an oven is used to electrically join the chip and the chip carrier. This process has a moderate auto-alignment capability due to the physical properties of the melted solder;

■ Gold bumps. They are directly put on the Al or Au bondpads of the chip. Different joining techniques exist:

  – Thermocompression. A combination of heat and pressure is used to make the electrical/mechanical bond. This is in fact the same technique as used to place bondwires. However, here the total applied pressure is proportional to the total number of bumps. Mostly, a bonding force of about 1N per bump is used and a temperature of around 200° is applied;

  – Thermosonic joining. The same procedure as the previous one, but additionally ultrasonic power is used to speed up the welding process;

  – Joining using adhesives. The adhesive can be non-conductive or conductive. In the latter case, a choice can be made between isotropic and anisotropic conductive adhesive;

For both thermocompression and thermosonic joining, the horizontal alignment is very critical as will become clearer in the next section.

To increase the mechanical strength of the bonded chip and to reduce mechanical stress due to the difference in thermal expansion coefficients of the chip material and the material of the chip carrier, mostly an underfill is used. It is dispensed alongside the bonded chip and it is drawn underneath the chip by capillary forces. After that, a medium to high temperature curing step is foreseen to complete the flip-chip joining process. Care must be taken when using underfill, since it can seriously influence the performance of circuits that rely on the electrical characteristics of on-chip passives like inductors [Carc PhD01, Wads ICM98].

Among the disadvantages of flip-chip bonding is the long curing time needed for the underfill to harden. However, multiple chips can be cured together, thus reducing the time needed per chip. Another disadvantage is the need for X-ray equipment to inspect hidden joints and the fact that no repairs can be made after the underfill has cured. As will be shown further on, care must be taken when X-rays are used to inspect a bonded chip since this method does not always leave the electrical properties of the chip unaltered.

2.5.2   Flip-Chip Bonding

The bonding technique that has been used for the upconvertor in Chapter 5 consists of a manual placement of gold bumps on the Al bondpads of the chip and a manual alignment and thermocompression joining step. In Fig. 2.35 a flip-chip bonded die on an alumina test substrate is shown. The conducting leads on the substrate are printed using a thick-film process. Underneath the chip, a Au conducting paste is used in combination with a 100μm-pitch process while for the rest of the substrate a Ag conducting paste is used in combination with a 300μm-pitch process. After the chip is bonded on the substrate it is placed in a CuBe container that shields the chip from RF-interference during measurements. Part of this container can also be seen on Fig. 2.35.

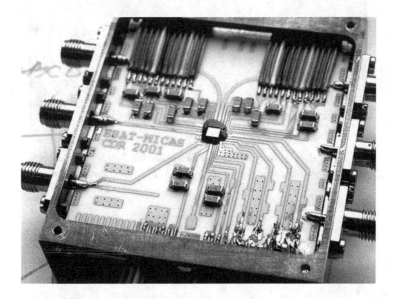

Figure 2.35. Flip-chip bonded die on alumina substrate in CuBe container

The following paragraph demonstrates the alignment problems that can occur during the ramp-up of a flip-chip bonding process using gold bumps. A second paragraph demonstrates the effect of the X-ray irradiation occuring during inspection of a flip-chip bonded sample on the chip's electrical behavior.

2.5.2.1 Alignment Problems

Since the alignment in the used flip-chip joining process is done manually using a semi-transparent mirror, it is not as such error prone. In Fig. 2.36 an X-ray picture is shown of a chip that is clearly misaligned with the leads on the substrate. As can be expected, this chip did not function properly.

In Fig. 2.37 a good sample is shown. Here, each bump coincides nicely with a lead on the substrate. This chip did work properly before the picture has been taken.

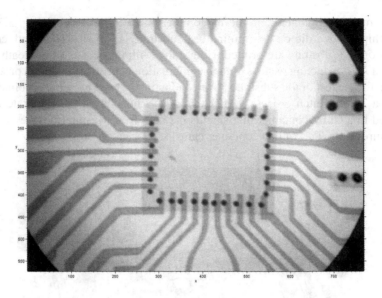

Figure 2.36.   X-ray picture of a misaligned flip-chip bonded die
X-ray pictures made at the department MTM, KULeuven, Belgium.

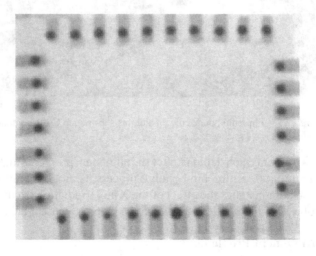

Figure 2.37.   X-ray picture of a well aligned flip-chip bonded die

2.5.2.2   Effect of X-ray Irradiation

As demonstrated in the previous subsection, correct alignment of the die on the substrate is not always a first-time right process. Therefore, inspection of the alignment of the bonded die is necessary. While for an automatic bonder the inspection is only needed during the setup of the bonding process, it is wishful to inspect every bonded die separately when a manual bonder is used as is the case here. A possible inspection method is the use of X-rays to check the alignment of the bumps with the leads on the chip. Since the bumps are placed manually on the bondpads of the chip, the alignment of bump and bondpad is inherently correct.

For testing purposes, a working sample of the two-channel transmitter described in Section 4.1.4.2 has been selected. The fact that it is wire-bonded doesn't have an influence on the results of this experiment. The chip contains following elements:

- Two on-chip ringoscillators with independently selectable oscillation frequencies;

- Each ringoscillator is used as LO for a transmit path;

- Each transmit path consists of a quadrature linear mixer and an RF current buffer.

The two transmit paths are called channel A and B. During the measurements described here, a quadrature baseband signal at 200 kHz has been applied to the baseband inputs of the mixers. Before irradiation, the measurement results shown in Table 2.5 have been obtained.

Table 2.5. Measurements on 2-channel transmitter before irradiation

	f_{osc}	P_{out}
Channel A	668.00 MHz	-19.9 dBm
Channel B	805.01 MHz	-20 dBm
BB_{IN}	860 mV	
BB_{SET}	1.97 V	

Here, f_{osc} is the LO frequency, P_{out} is the power at $f_{osc} + 200 \, kHz$ and BB_{IN} is the DC voltage measured at the baseband inputs of the mixers. From the description of the transmitter circuit in Section 4.1.4.2 it will become clear that this voltage is set by a simple cascode transistor. The gate voltage of this cascode transistor is given in Table 2.5 as BB_{SET}.

To align the CuBe container, first a low-energy X-ray beam ($< 80kV$) has been used during a time of about 2 minutes. To be able to see the chip, a higher energy of about 125kV is needed with a dose of 0.3 mA. This beam has been used for 2 minutes, after which it has been shut down. Then, the container has been removed and 10 minutes later the measurements shown in Table 2.6 have been registered.

Table 2.6.   Measurements on 2-channel transmitter 10' after irradiation

	f_{osc}	P_{out}
Channel A	664.20 MHz	-17.9 dBm
Channel B	809.2 MHz	-20.02 dBm
BB_{IN}	912 mV	
BB_{SET}	1.97 V	

As can be seen, the oscillation frequency is shifted to a lower frequency for channel A and to a higher frequency for channel B. The output power of channel A has increased with 2 dB while the output power of channel B is almost the same. Since the DC voltage at the baseband inputs of the mixers is applied by a cascode transistor with a known gate voltage, any shift in V_T can be observed directly in the measurements. Thus, a V_T shift of 52 mV has been measured.

After 2 hours, the measurements given in Table 2.7 have been made.

Table 2.7.   Measurements on 2-channel transmitter 2h after irradiation

	f_{osc}	P_{out}
Channel A	662.40 MHz	-17.6 dBm
Channel B	807.7 MHz	-22 dBm
BB_{IN}	908 mV	
BB_{SET}	1.97 V	

Looking at the oscillation frequency and the V_T shift, one could conclude that some curing has occured and that after a longer waiting period the effects would possibly disappear completely. However, looking at the output power of both channel A and B one can only conclude that the irradiation incurs inpredictable long-term effects on the chip. The conclusion here is that during the ramp-up of a flip-chip process, the influence of the X-ray inspection must not be overlooked. In the context of this work, it has not been investigated if a combination of (lower) X-ray energy and dose exist that does not result in measurable effects on the electrical properties of the chip. This would be the goal of a production line optimization of the inspection procedure.

2.6   Conclusions & Use of Presented Topics

In this chapter a large variation of topics has been handled. The discussion in the first section about substrate resistivity fully supports the choice of a technology with a high substrate resistivity for the successful RF designs presented in Section 4.2 and Chapter 5. As will become clear further on, the use of low-resistivity substrate for the ringoscillator design presented in Section 4.1.4 has not deteriorated the performance since no on-chip inductors are included in that particular design. However, the design of a 3.3 GHz VCO presented in Section 3.4.6 does show a lower performance than the RF VCO described in Section 4.2 due to the use of a low-resistivity substrate.

The inductor model presented in the second section and the method used to obtain it with FastHenry simulations, has been employed for all inductors designed in this work. No measurement-based inductor libraries have been used. The RF MOST model has been used for the design of the upconvertor presented in Chapter 5 and in a simplified version it is used for the design of the RF VCO discussed in Section 4.2.

Quadrature local oscillator generation is needed in the upconvertor design described in Chapter 5, for which a choice had to be made between the two methods presented here. An alternative based on direct quadrature generation using a ringoscillator, is used in the transmitter for cable applications presented in Section 4.1.4.

Finally, flip-chip bonding is used in the manufacturing process of the upconvertor (Chapter 5). During the design, the need for low bondwire inductance enforced the migration towards this bonding method. The use of it, however, did introduce a serious delay in the measurement process due to the total absence of a possibility to make corrections after the chip has been bonded.

2.6 Conclusions & Use of Presented Topics

Chapter 3

AUTOMATED VCO SYNTHESIS

3.1   Introduction

Where digital designers can rely for some years already on the powerful productivity enhancing functionality of a variety of design tools, the research for tools that can provide the analog designer with a comparable enhancement in productivity still has some challenges to tackle. This chapter describes a methodology and a practical implementation of tools for analog design, based on the combination of expert knowledge and global optimization. The conceptual set-up envisaged for this is not to provide for a framework that enables an unexperienced user to start making very complex analog designs at high or low operating frequencies. On the contrary, the methodology that will be proposed expects a good knowledge of the user, and enables expert designers to build their own library of re-usable designs that can possibly be shared with other designers. In this context, re-usability is to be interpreted as technology independence and, within the limits of the used technology and the topology described in the design-specific library entry, specification independence.

Intentionally, no use has been made of the "CAD"-word (Computer-Aided Design) in the title of this chapter, since this would have been a little misguiding. Compared to the generic high-level approach most CAD-related software, literature and people show to be adherent of, the tools for automated analog building block design described here might seem ad-hoc and possibly non-reusable. The ad-hoc property is effectively one of the core elements of the methodology and stems from the direct integration of expert design knowledge about a single circuit into a "template" from which a circuit can be generated for a certain set of specifications and a certain technology. This can be done in a single run or a single optimization process comprising multiple execution steps. In this context, an acronym that would be less misguiding than CAD would be DAC: *Designer-Assisted Computing*, emphasizing the inherent active job of the designer who leads the computer through a predefined systematic work flow. Within this work flow, the computer performs those jobs it is suited for: dumb, brute-force calculations and simulations, leaving for the designer the job he's being paid for: being creative in the pursuit for higher performance, lower power, smaller area, ... analog circuitry. This view of CAD-methodology is certainly in contradiction with the high-level design and push-the-button implementation concept that some still adhere to, but that has been proven even for digital not to be the most successful way to go for cutting-edge performance designs.

Since this work is about analog RF design, this chapter will focus on a methodology for analog circuits. It is the sole intention to describe a practical and workable approach towards more resource-efficient analog circuit design. The problems of digital/analog partitioning, co-simulation and co-design will therefore not be treated.

The first section defines and situates the proposed methodology for a systematic analog design environment within the field of existing methodologies for structured analog design, and introduces the definition for a "functional block" as used in the presented work. In a second section, the elements of a general optimization process are introduced and a design-template based implementation of automated optimized design of functional blocks is proposed. In a third section the CYCLONE tool ([DeRa DAC2000]), that enables the design of LC-type VCO's, is described in detail as an example of the proposed methodology and design-template based implementation. In a final section some conclusions are formulated.

3.2   Structured Analog Design Methodology

3.2.1   Definition of Functional Blocks

The goal of any design methodology should be to obtain "optimally" designed circuits:

- Chips that meet the necessary specifications without major over-performance, for the lowest area/power/cost[1];

- During circuit (re-)design, optimal use is made of available resources: people, design time and available know-how within the company.

---

[1]The production yield certainly influences the cost, but has a direct influence only on the *marginal* cost of one chip. It is beyond the scope of this text to make a detailed study of the economics involved in chip design and fabrication, but due to the rising cost per area for the newest deep-submicron CMOS technologies, the choice of the right technology for the right kind of circuit will become more and more of an issue to render a project successful.

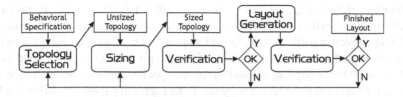

Figure 3.1. Analog design flow

A general design flow for an analog circuit design, is given in Fig. 3.1 [Donn PhD98]. This flow is valid for all analog circuitry that can be regarded as a single entity to be designed as a whole. Such entities can be small blocks like opamps, but the flow can also be used for large systems with distinctive subsystems. Depending on the source in literature, a lot of different decomposition schemes exist that can be used to divide a large analog system in different subsystems. Since these considerations mostly stem from a pursuit towards generic high-level automation and since this is not at all the intention of this work they will not be discussed here. Moreover, since a unified approach for this decomposition does not exist, and probably never will, it does not bring much added value to the discussion. However, we do need a tangible object that we can use as basis for the structured design flow that is proposed here. Therefore, the *functional block* as used in this work is introduced.

Functional Block   Let us just focus on the very basics, and call a subsystem in this work a *functional block* if it has following properties:

- It has a fixed topology;
- It's function within the total system can be described in either mathematical form or natural language;
- It's functionality and performance can be quantified using a limited number of parameters as GBW, gain, bandwidth, power, input/output impedance, ...;
- It's coupling with other building blocks is well defined.

The last property means that the use of the block within a system will not lead to a dramatic change in behavior of the block itself and of all blocks it is connected to. In fact, this property introduces the need for a proper partioning of an architecture into functional blocks. It is assumed here that this is done by the expert designer. *E.g. a bad partioning would be to isolate a single-transistor feedback from a more complex circuit.*

In fact, we now defined the *hierarchical level* at which we want to implement analog design automation and use optimization. We did not define a *complexity level* for the functional block. This means that a functional block can have a circuit complexity starting from a single transistor going up to e.g. a complete RF upconvertor path as presented in Chapter 5. The reason no complexity level is mentioned is that the complexity of the circuit that has to be handled as one functional block depends strongly upon the specific application, the frequencies involved and even the technology used. To exemplify this: 500 MHz is a high frequency in a 0.7μm CMOS technology, but it certainly is not in a 0.12μm CMOS technology.

In the next section, the defined functional block is situated within a broader view on structured analog design. What should be clear already is that the definition of functional block is kept quite vague, perhaps incomplete, and is certainly more content-driven than formalization-driven. This makes the defined structure probably useless for formal descriptions of circuits, that might ultimately lead to a generic structured, thus programmable, approach for analog design and synthesis. However, the approach used here results in a practical and designer-friendly environment that invites the designer to opt for a formalized design flow that inherently opens the path to re-use and low-level optimization. If the methodology is used in the right way, an optimized redesign of a complete system can be done in a fraction of the original design time, using an interactive approach in which the designer keeps control of all design aspects and that possibly even incorporates layout-related issues during design.

3.2.2   Structured Analog Design

The ultimate dream that CAD-software could offer to a designer, is of course a high-level design environment in which complete abstraction can be made from all implementation-related aspects of the final circuit. A first effort to realize this dream, and that is nowadays recognized as a usable instrument to describe a structured design approach, is the well-known hierarchical top-down/bottom-up design methodology, of which a two-level version is represented in Fig. 3.2. Starting from the high-level circuit on the top, the goal of the methodology is to obtain at the highest level the layout information of the implementation levels below to verify if the obtained behavior of the physical layout agrees with the wanted behavior. In the first level the circuit is synthesized at subsystem level, and some crude layout information becomes already available. The specifications of the introduced subsystems are then used in the next lower level for a synthetization at block-level. The complexity of each element in the compound circuit thus decreases from top to bottom, while the layout detail increases concurrently. At the bottom level, the output of the synthetization is a circuit at device level and the physical layout. The information about the layout at top level can then be used to do the final performance verification.

An approach to implement this structured design methodology in an automated way has been proposed by [Donn PhD98]. It splits the analog design

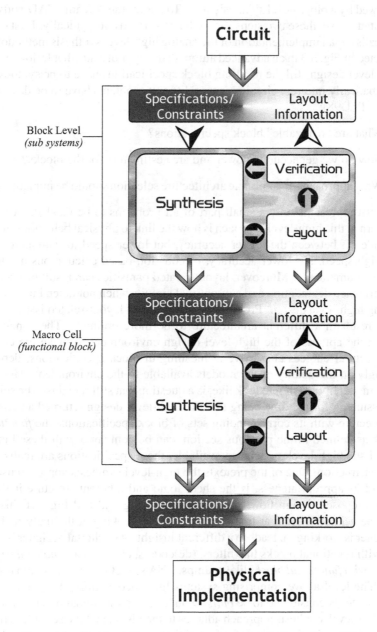

Figure 3.2. Two-level hierarchical design methodology [Levin Spect02]

into two parts: a high-level synthesis step resulting in a block-level schematic, followed by a block-level synthesis step. To realize the "dream" CAD-software as stated above, these two steps should be performed automatically. First of all, this calls for an implementation of the analog high-level synthesis methodology depicted in Fig. 3.3 in an advanced automated way. Put aside that following this high-level design still the resulting block specifications have to be synthesized automatically into real circuitry, some difficult problems have to be dealt with [Donn PhD98]:

■ What are "realizable" block specifications?

■ How do we get accurate power and area estimators for the blocks?

■ What approach for automatic architecture selection should be implemented?

This list represents only a small part of all problems to be tackled. A major concern in this high-level approach is how the link to physical behavior is made. A trade-off between the higher accuracy, but lower speed, of simulation and the high speed, but lower accuracy, of equation-based calculations has to be made automatically. Moreover, layout-related parasitics can result in a serious performance degradation and sub-optimal circuits when not taken into account during high-level design. Possibly the inclusion of layout-related issues might even result in a different architecture to be more optimal. The conclusion is that the aptness of the high-level design environment to make the correct architectural choices and arrive at meaningful block specifications, depends strongly on the quality of the models available to the environment. How the ideal model library should look like, is a question that still remains to be solved.

Assuming for the time being that the high-level design returned an eligible architecture with its corresponding sets of block specifications, the *functional block* as defined in the previous section, can be compared with these blocks. If realistic high-level models are available, these specifications are realizable-by-construction. One of the pretexts the high-level synthesis approach uses to defend its appropriateness, is the sheer logic and inherent structure it shows in its proposed design flow. Looking at both Fig. 3.3 and Fig. 3.1 this can only be acknowledged, at first sight. However, looking at the daily reality a designer is working in, leads to a different insight. As a digital designer is dealing with functional blocks like filters, (de)coders, etc., also an *analog designer works with functional blocks* like opamps, LNA's, VCO's, mixers, comparators, etc. The level above comprising systems like upconvertors, downconvertors, ADCs, etc. is considered to be *built up* out of these functional blocks, whereas the high-level synthesis approach adheres to the vision that the set of functional blocks is merely an *implementation* of the behavior of the top system level. Although the latter vision on circuit synthesis showed very valuable in digital logic design synthesis because in every phase of the synthesis process an in-

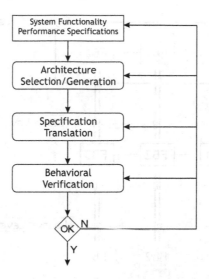

Figure 3.3. Analog high level synthesis

trinsic correctness-by-construction can be guaranteed, it falls short for analog design synthesis because there the exact logical (digital) equality of behavioral function and its implementation is replaced by an "equivalence" of analog behavior and its implementation, since any "real" analog building block inherently introduce errors due to noise, mismatch etc.

3.2.3 Design-Level Entry Methodology

To alleviate the shortcomings of the strictly structured top-down approach as described in the previous section, a methodology is proposed that tries to incorporate following issues that could be regarded as the properties of an "ideal" tool for an (expert) analog designer:

- The designer can stick to his normal design practice;

- Expert knowledge is captured at the functional block level;

- The functional block level is the working level of the tool, with possibly a high-level design layer above and a layout level beneath it;

- Layout parasitics can be taken into account;

- The description of the functional blocks is technology independent.

The *"Design-Level Entry'* methodology (DLE) is based on these properties and tries to interfere as little as possible with the way the analog designer works in day-to-day practice. Prior to the real design work, a general system study and

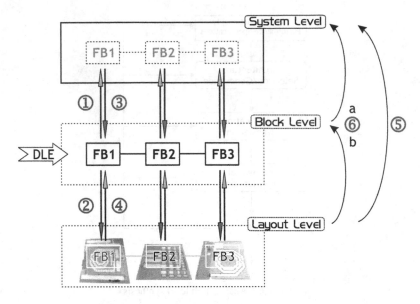

Figure 3.4.   The Design-Level Entry Methodology

the choice of an architecture to start with is done, based on system expertise.[2]
The big difference with the high-level synthesis approach in this stage of the
design flow is that the decomposition of the system is not based on block models,
but on know-how found in-house or in literature.

In Fig. 3.4 the Design-Level Entry methodology is visualized. Numbered 1
to 4 are the interactions that exist between the Block Level, System Level and
Layout or implementational level. Starting from the assembly of *functional
blocks* out of which the designer will build the top level circuit, he constructs
and refines the high-level model of the circuit (1). He also sets up a *design plan*
for each functional block to arrive at a physical implementation of each block
(2). Feedback from the layout implementation is used in the functional block
simulations (4). The specifications of the functional blocks are derived and
fine-tuned using the high-level simulations (3). If possible, final verification
for the whole circuit is done at device level (5). Otherwise, the refined high-
level model is used for final verification (6a), combined with verification of the
individual functional blocks at device level (6b).

The *high-level synthesis methodology* tries to enable less expert analog designers to start designing complex systems. It therefore hides as much as possible the expert knowledge about circuit implementation in a large database

---

[2]In fact, this can also be seen as "manual" synthesis in the top level of the hierarchical design flow of Fig. 3.2.

consisting of circuit and device models. Even for the synthesis of the blocks in lower levels of the hierarchical design flow, the user is shielded from the real implementation issues involving a high expertise. On the other hand, the everyday practice shows that the knowledge as applied by the expert designer is linked directly to the functional block level, as explained in the previous paragraph. Thus, a fundamental difference exists in the hierarchical level a designer is working at in day-to-day work and the hierarchical level high-level synthesis is expecting the designer to be working at. Whereas this should not be a big issue for a novice designer, an expert designer will always want a transparent view on what is happening, and if necessary he wants to be able to steer the synthetization process according to his expertise. A closed synthetization system that shields all knowledge from his users and that doesn't allow the incorporation of the knowledge of an expert user, will probably not be used by this expert designer.

DLE Tool   To implement the methodology described above, a tool is needed that helps the designer to:

- Automatically generate a high-level model of each chosen functional block *(Block Modeling Function)*;

- Generate a sized block including layout *(Block Sizing Function)*.

Such a tool does not force the designer to dramatically change his everyday design practice, and it enables the same designer to leave all boring aspects of his daily job to the computer. Moreover, the tool has to allow the designer to implement his knowledge into the design of the functional block itself. After the design of a functional block two templates remain: a *Design Template* that includes the design know-how, and an *Evaluation Template* that describes the test benches needed for the evaluation of the functional block. These Templates could be shared among (expert) users within a company or sold as *analog IP (Intellectual Property)* to other companies. This is further elaborated in the next section.

Convergence between methodologies   It's quite obvious that both the high-level design methodology and the proposed DLE methodology stem from a common global view on structured analog design. Therefore, a convergence between the two methodologies is to be expected at a certain point. This convergence occurs when the complexity of the functional block used in the DLE methodology grows to the level of a complete system architecture, being the toplevel of the high-level design methodology of Fig. 3.2 or Fig. 3.3. However, the difference that still remains is that the DLE methodology results in a fixed architecture with an inherently linked design plan set up by the expert designer, whereas the use of the high-level design methodology does not automatically results in a certain design plan. Another common point is that both methodologies can deliver an optimized design when linked to an optimization algorithm. However, the use of a design plan in the DLE enables an active reduction by the designer of the number of variables of the optimization process as compared to a fully automated high-level design synthesis.

It can be concluded that the DLE methodology coincides to a large extent with the high-level design methodology. Nevertheless, two distinctive properties render the introduction and use of the proposed DLE methodology useful:

- The point-of-entry in the methodology is at the level of the expert designer in his own habitat, being the design level of functional blocks. This makes a smooth transition to the proposed methodology possible;

- The availability of a design plan after a single iteration of the DLE leads to a more efficient use of an optional optimization algorithm, when redesign in another technology or global design optimization is needed.

3.3   Automation in Functional-Block Design

3.3.1   Introduction

In a first subsection, the elements of a general optimization process are introduced, and the concept of knowledge-based performance-driven optimization is explained. It is shown that this view on optimization enables the inclusion of design knowledge within a specific part of the optimization process, being the cost function construction/evaluation. In the second subsection, the template-based design of functional blocks is proposed as an implementation of the *Block Sizing Function* of the DLE Tool described at the end of the previous section. The combination of this template-based design method with the knowledge-based optimization process eventually leads to a template-based design optimization flow that can be used to speed up the everyday design work. In this work the *Block Modeling Function* of the DLE Tool as presented above has not been implemented. Commercially available software as [Xpedion] prove that this function can be integrated within the known design environment.

3.3.2   Optimization Process and Algorithms

An optimizer can be very useful to aid a designer in his pursuit of the most optimal circuit design or in his effort to meet design or re-design deadlines. The algorithm that is used within the optimizer is a critical part within the structured analog design flow as is proposed in the next subsection, since it determines the way the independent design parameters are changed in order to obtain the most optimal design. In a first subsection, the elements of a general optimization process are introduced. A second subsection discusses the construction of a cost function. The last part deals with optimization algorithms that can be used in circuit-design optimization.

3.3.2.1 Optimization Process

In Fig. 3.5, a general optimization process is shown, with the optimizer as central block and with a clear representation of all elements of the optimization process:

- The optimization parameters; Consisting of *independent* and dependent parameters. They can also be classified as input parameters, *intermediate* parameters, *constant* parameters and output or *performance* parameters.

- Intermediate processes as calculation, circuit sizing and circuit evaluation;

- Cost function; Its construction is discussed in the next subsection.

The optimization process is initiated by the optimizer that generates a set of *independent* parameters (1). These are used in a *calculation* process (2) together with some *constant* parameters (3) to produce a set of *intermediate (dependent)* parameters (4). These intermediate parameters are for instance sizes of transistors that are designed to have a certain transconductance (gm) and gate-overdrive voltage ($V_{gs} - V_T$), but could as well be the electrical parameters of an inductor with a geometry defined by a set of constant and/or independent parameters that is simulated using an external coil simulator program. Once all parameters are calculated, the sized circuit (5) is fully defined by a combination of independent, constant and intermediate parameters. Then the circuit can be evaluated (6) using a circuit simulator as HSpice or SpectreRF. This results in a set of *performance* parameters (7). These are used to evaluate a *cost function* (8), resulting in a certain value for the *cost* (9). This cost is furnished as input to the optimizer that then generates a new set of values for the independent parameters. The optimizer is used to seek a minimum of the cost function by exploration of the design space with a dimension defined by the number of *independent* parameters.

The process as described above consists out of two parts, clearly indicated in Fig. 3.5:

- A "standard" optimization process; It uses an optimizer to minimize a cost function by exploration of the design space defined by a set of independent parameters;

- A knowledge-based performance calculation; This could be regarded as a "black box" implementation of a customized cost function that, given a set of values for the independent parameters, returns a certain cost.

The combination of these two processes can be given one descriptive name: a *Knowledge-Based Performance-Driven Optimization Process*.

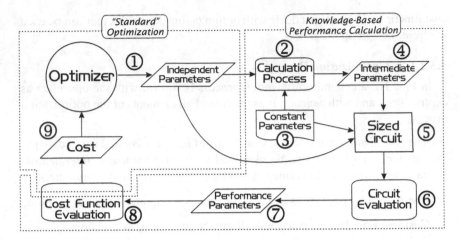

Figure 3.5.   General circuit-optimization process

3.3.2.2   Cost Function Construction

In this work, the overall goal of the optimization is to obtain a sized circuit that meets a set of p performance specifications referred to as $S_1 \ldots S_i \ldots S_p$. Different ways to formulate these specifications exist:

- Equalities;

- In-equalities;

- Properties that have to be minimized/maximized.

   Exemplary set of specifications
   S_1 :   Oscillation frequency= 3 GHz
   S_2 :   Tuning range > 20%
   S_3 :   Phase Noise < -100 dBc/Hz
   S_4 :   Power consumption = minimal
   S_5 :   Chip area = minimal

Based upon the set $\{S_i\}$ of design specifications, a cost function is constructed that expresses in a single value how good all design specifications are met by a circuit sized starting from a specific set of independent parameter values. The cost function is formulated as:

$$Cost = \sum_{1 \le i \le p} w_i C^i_{rel} \qquad (3.1)$$

with w_i the weight factor and C^i_{rel} the relative cost contribution of specification S_i, describing how well the performance parameters of the evaluated design meet specification S_i. The weight factor determines the importance of the corresponding design specification within the total optimization process. Properties that have to be minimized can be used directly as relative cost. Those to be maximized can be added likewise after inverting them. The relative cost of specifications formulated as an (in)equality is defined in such a way that it has a minimum around the value for which the specification is met.

In general, a cost function constructed in such manner will have a quite irregular shape over the design space. Moreover, some of the independent design parameters have a continuous character, whereas others are strictly discrete. The optimization algorithm that is used to seek a global minimum of the cost function therefore has to be quite robust and should not be based on any assumptions about the behavior of the cost function or its derivatives. Two candidates for such an algorithm are briefly discussed in the next subsection.

3.3.3 Optimization Algorithms

The two algorithms that will be discussed here have been used successfully in tools for automated design optimization. First, simulated annealing and its different implementations will be presented. In the second part, the genetic algorithms and more specific genetic evolution will be handled. Recently, geometric programming has been shown to be quite useful for design optimization [Hers TCAD01]. The major advantage is the convex property of the optimization problem, enabling the use of very fast and efficient optimization algorithms. However, the drawback is that only problems and that can be represented in a particular way can be dealt with. Moreover, each new circuit topology has to be converted into a "geometric program". Therefore, this method of optimization can not be implemented as such in the framework as described in the previous section and will not be discussed here any further.

3.3.3.1 Simulated Annealing

Simulated annealing is a generalization of a Monte Carlo method for examining the equations of state and frozen states of n-body systems [Metr JChem53]. The concept is based on the manner in which liquids freeze or metals recrystallize in the process of annealing. In an annealing process a melt, initially at high temperature and disordered, is slowly cooled so that the system at any time is approximately in thermodynamic equilibrium. As cooling proceeds, the system becomes more ordered and approaches a "frozen" ground state at $T = 0$. Hence the process can be thought of as an adiabatic approach towards the lowest energy state. If the initial temperature of the system is too low or cooling is done insufficiently slowly the system may become quenched forming defects or freezing out in meta-stable states (i.e. get trapped in a local minimum energy state).

The original Metropolis scheme is that an initial state of a thermodynamic system is chosen at energy E and temperature T. Holding T constant the initial configuration is perturbed and the change in energy dE is computed. If the change in energy is negative the new configuration is accepted. If the change in energy is positive it is accepted with a probability given by the Boltzmann factor $e^{-(dE/T)}$. This processes is then repeated sufficient times to give good sampling statistics for the current temperature, and then the temperature is decremented and the entire process repeated until a frozen state is achieved at T=0.

The generalization of this Monte Carlo approach to fitting non-convex cost functions arising in a variety of combinatorial problems is straight forward [Kirk Science83, Cerny 85]. The current state of the thermodynamic system is analogous to the current solution to the combinatorial problem, the energy equation for the thermodynamic system is analogous to the cost function for the problem, and the final ground state is analogous to the global minimum. The major difficulty in the implementation of the algorithm is that there is no obvious analogy for the temperature T with respect to a free parameter in the combinatorial problem. Furthermore, avoidance of entrapment in local minima (also called quenching) is dependent on the "annealing schedule", the choice of initial temperature, how many iterations are performed at each temperature, and how much the temperature is decremented at each step as cooling proceeds [Gray 97].

The speed of annealing is in first order determined by the annealing schedule used to lower the temperature for the consecutive annealing steps [Ingb CModel93]. For the original "Boltzmann Annealing" it is proven that a statistically guaranteed global minimum of the cost function can be found [Geman 84] if the starting temperature T_0 is "high enough" and the temperature in each annealing step k is determined by:

$$T(k) = \frac{T_0}{\ln k} \qquad\qquad (3.2)$$

To speed up the sometimes very lengthy simulated annealing process, alternative annealing schedules are used. If implemented in a correct manner, this "Simulated Quenching" technique effectively results in a faster annealing schedule without losing the property of the guaranteed finding of the global minimum [Ingb CModel92].

3.3.3.2   Evolutionary Algorithms

Evolutionary algorithms (EAs) are search methods that take their inspiration from natural selection and survival of the fittest in the biological world [Gray 97]. EAs differ from more traditional optimization techniques in that they involve a search from a "population" of solutions, not from a single point. Each iteration of an EA involves a competitive selection that weeds out poor solutions. The solutions with high "fitness" are "recombined" with other solutions by swapping parts of a solution with another. Solutions are also "mutated" by making a small change to a single element of the solution. Recombination and mutation are used to generate new solutions that are biased towards regions of the space for which good solutions have already been seen.

Several different types of evolutionary search methods have been developed independently. These include (a) genetic programming (GP), which evolve programs, (b) evolutionary programming (EP), which focuses on optimizing continuous functions without recombination, (c) evolutionary strategies (ES), which focuses on optimizing continuous functions with recombination, and (d) genetic algorithms (GAs), which focuses on optimizing general combinatorial problems.

EAs are often viewed as a global optimization method although convergence to a global optimum is only guaranteed in a weak probabilistic sense. However, one of the strengths of EAs is that they perform well on "noisy" functions where there may be multiple local optima. EAs tend not to get "stuck" on a local minima and can often find a globally optimal solution.

The recombination operation used by EAs requires that the problem can be represented in a manner that makes combinations of two solutions likely to generate interesting solutions. Consequently selecting an appropriate representation is a challenging aspect of applying these methods.

The use of genetic algorithms for circuit optimization showed its merits in tools for opamp synthesis like DARWIN [Kruis DAC95] and tools for high-level design of sigma-delta convertors as DAISY [Fran DATE02]. It is to be noted that in these tools one of the major tasks of the optimization algorithm is to select or even create the best topology for a certain given set of specifications. It could be concluded that genetic algorithms have some advantages for problems involving topology selection. However, also simulated annealing has been used successfully to solve the topology-selection problem ([vdPlas TCAD01]).

3.3.3.3 The ASA Algorithm

ASA (Adaptive Simulated Annealing) [ASA] is developed to statistically find the best global fit of a nonlinear non-convex cost function over a D-dimensional space. This algorithm permits an annealing schedule for 'temperature' T decreasing exponentially in annealing-time k, $T = T_0 e^{(-ck^{1/D})}$. The introduction of re-annealing after a certain number of temperature decrements also permits adaptation to changing sensitivities in the multi-dimensional parameter space. This annealing schedule is faster than fast Cauchy annealing, where $T = \frac{T_0}{k}$, and much faster than Boltzmann annealing, where $T = \frac{T_0}{\ln k}$. For "new" problems of which the properties of the cost function are unknown or not known well enough to tailor a certain (local) optimization scheme to the problem at hand, it has been shown that the use of ASA results in significant better results than other optimization algorithms. Compared to fast, local optimization algorithms, ASA has the advantage of a certified global minimization, and comparison with genetic algorithms showed ASA needs a lower number of cost function evaluations to arrive at the global minimum [Ingb CModel92]. For the problems at hand in this work, the use of ASA therefore is the most appropriate.

The major reasons for which the simulated annealing algorithm in its ASA implementation is used in this work are summarized here:

- The statistically guaranteed finding of an optimal solution, at an acceptable speed due to the very fast annealing schedule and the adaptive reannealing;

- The ability to handle arbitrary complex cost functions, with any number and type of nonlinearities and discontinuities;

- The ability to deal with arbitrary boundary conditions and constraints;

- The possibility to combine both integer and real parameters;

- The ease of use and implementation within a complex framework.

3.3.4   Template-Based Design of Functional Blocks

The base of all analog design is a device schematic. Therefore, the starting point of any design tool should be the schematic of the circuit under consideration. In this section, the description of a tool is proposed that can be used to realize the Block Sizing Function of the DLE Tool described in Section 3.2.3. Combining it with the knowledge-based optimization process introduced in Section 3.3.2, leads to a template-based design optimization tool. Apart from the optimization which will be added at the end of this discussion, such a tool implements following functionalities:

1 Schematic entry of the circuit as starting point;

2 Circuit Design Directives for the sizing of a full functional block can be added by the designer;

3 Automatic Device Sizing capability;

4 Automated performance estimation, based on Evaluation Templates;

5 Automatic layout generation, including parasitics estimation on both device level and chip level (interconnect, bondpads, ...);

As shown in Fig. 3.6, the Circuit Design Directives (2) include the know-how of the designer and together with the circuit schematic (1) they make up the *Design Template* that fully describes the design of the functional block. From implementational side of view, a good way to avoid unnecessary difficulties is to work only with fully parameterized schematics as input for the tool. Clearly, no topological changes are made automatically by the tool. This is the job of the expert designer using the tool.

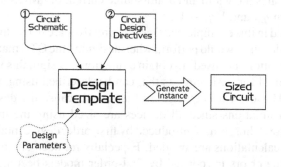

Figure 3.6. Circuit instance generation based on Design Template

Consider an implementation of a design tool that realizes items 1 to 3 of the list above, being a Design Template and Automatic Device Sizing. Such a tool in fact implements an equation-based design methodology with circuit-simulation accuracy. The equations are the Circuit Design Directives as entered by the designer during the built up of the Design Template, whereas the accuracy can be ensured by the Device Sizer. Fig. 3.6 clarifies the construction of a parameterized Design Template and shows the generation of a sized circuit instance from this Design Template. The tool directly enables the automated *design* of a few often-used functional blocks such as simple opamps and the automatic *generation* of repetitive circuits like polyphase networks[3] and more complex circuits like RF current buffers such as in Chapter 5.

The automatic design of the symmetrical OTA shown in Fig. 3.7 can be realized with this simple tool using a Design Template consisting of the parameterized circuit schematic of Fig. 3.7 and Circuit Design Directives of which some are given in Code Ex. 3.1. The directives shown have been written to obtain an OTA with maximum *GBW* given a minimum phase margin (*PM*) for a certain g_m of the input pair and mirror factor *B* [Lake 94]. The maximum *GBW* that still respects the minimum *PM* is calculated using an iterative process. An initial value is given by setting $GBW = tan(PMmin) * fnd1$, with *fnd1* the first none-dominant pole. A better estimate for the *PM* is then calculated and compared to the wanted *PM*. If the margin is too low, the *GBW* is lowered and if it is too high, the *GBW* can be raised[4]. After determination of the maximal

---

[3]Here, one of the design parameters is the number of stages of the polyphase network
[4]In a real implementation, it would be better from numerical point of view to set a small *interval* of acceptable values for the *PM*.

GBW, the minimal capacitive load can be calculated that will lead to a realization of the calculated *GBW*. The code example is quite self-explanatory; only the lines where the transistor sizes are calculated need some explanation: the first line calculates the width and drain-source current of nMOS transistor *M1*, that has a given g_m and $Vgs - Vt$.

As is clarified in this example, the Design Directives are in fact the actions and calculations a designer would perform when designing the OTA manually. Since no circuit simulations are used[5] to obtain a functional design, this version of the tool is only suited for basic circuits that can be described using mathematical expressions. However, it is quite useful already because the designer avoids several design iterations since all devices are sized using the models of the circuit simulator. Thus, errors introduced by first-order device models normally used for hand calculations are avoided. Especially for deep-submicron CMOS technologies the errors introduced by first-order models become substantial. Conclusion is that this version of the tool can be regarded as an extensive, accurate device calculator that performs calculations as indicated in the Circuit Design Directives.

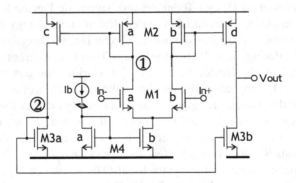

Figure 3.7.   Simple symmetrical OTA

For circuits of which the behavior and performance can not be described by simple formulae, circuit simulation and evaluation of the output behavior must be performed. This calls for the implementation of the *Evaluation Template* of the functionality-list given above (item 4). Fig. 3.8 shows the process of instance generation based on the Design Template and consequent circuit simulation based on the Evaluation Template resulting in a set of values quantifying the circuit instance behavior and performance.

---

[5]The sizing of the individual transistors is not seen as a circuit simulation. However, the implementation of the Sizing Tool could necessitate single-transistor simulations to obtain a high accuracy.

Code Example 3.1 Circuit Design Directives for OTA of Fig. 3.7

```
*Symmetrical OTA Design Directives

**Input Values**
gmin = 2e-3  # gm of the input pair
vgstin= 0.25  # vgst of the input pair
B = 1  # mirror factor
PMmin = 70 # wanted Phase Margin
length1= 0.25um
...

**Transistor Sizing**
calculatesize M1 ''width ids'' gm=gmin vgst=vgstin type=nmos
calculatesize M2 ''width gm'' ids=M1(ids) type=pmos ···
...

**Calculations**
Cnode1 = Cgs(M2a) + Cgs(M2c) + ...
fnd1= gm(M2) / (2*pi*Cnode1)
fnd2= gm(M3a) / (2*pi*Cnode2)
fz2= 2*fnd2
GBWmax= tan(PMmin) * fnd1
GBW= GBWmax  # start value for GBW
#Iterative calculation of GBW...
1>pha1= arctan(GBW/fnd1)
2>pha2= arctan(GBW/fnd2) - arctan(GBW/2*fnd2)
3>PM= 90 - pha1 - pha2
4>while PM <> PMmin: decrease GBW; goto 1>
Cminload = B*gm(M1) / (2*pi*GBW)
```

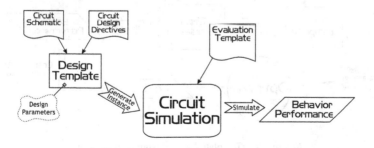

Figure 3.8.   Template-based instance generation and circuit simulation

In the design of critical blocks or blocks working at high frequencies, layout parasitics on both device level and chip level must be taken into account. These layout issues must be incorporated within the Design Template. An example of how this can be realized is given for a VCO in Section 3.4.5.

Now, only one element to arrive at a fully automated circuit optimization is lacking, being the optimizer itself. In Fig. 3.9 a complete overview is given of the automated template-based design-optimization flow that is proposed here as a usable tool for the design of optimized functional blocks. It combines the template-based design as described above with the knowledge-based optimization process introduced in the previous subsection. As is explained there, the values quantifying the behavior and performance of the generated circuit instance (the *performance parameters*) are used to evaluate a customized cost function.

The optimizer searches for a global minimum of this cost function by exploration of the design space defined by the user-specified independent parameters in the Design Template. The set of values for the independent parameters that minimizes the cost function, together with the values of the constant parameters and the values for the intermediate parameters calculated from those two parameter-value sets, describe the circuit instantiation of the Design Template that fulfills the Design Specifications in the most optimal way. The next section describes in detail how the method is used in practice for an important functional block in RF systems, being a Voltage Controlled Oscillator (VCO). In the remainder of this section, the use of weight factors is briefly discussed and the proposed template-based design flow is put in a historical perspective.

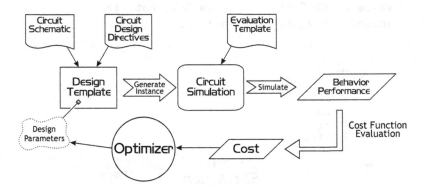

Figure 3.9. Template-based Design Optimization

Use of weight factor vs. direct optimization   An often heard objection against the use of weight factors in design automation is that it would be even less transparent to the user than a direct (manual or automated) optimization using the transistor parameters as design parameters. Of course, a good choice of weight factors is important but not the absolute value as such is important. It is the relative weight that will influence the final result. So indeed, the user will probably have to experience with the exact value of the weight factors the first time he uses a certain Design Template.

However, the big advantage of the knowledge-based performance-driven optimization process introduced in Section 3.3.2 is that the way the weight factors influence the result is independent of technology and for a certain class of circuits (e.g. oscillators) also independent of the chosen topology. Due to the fact that the Design Directives already guarantee a working design for a certain set of independent parameters, the optimization process does not have to find the boundaries of the valid design space.

Summarizing, the (relative) weight factors can truly be used to force the design performance to the desired corner of the design space. Once a Design Template has been constructed, the device sizing details are shielded from the designer that tries to design a certain building block towards a given set of performance targets. The designer can try out Design Templates from e.g. different topologies of a certain type of building block, using a design methodology based on changing the relative weight factors, which is independent of the specific topology.

Historical perspective   Comparing the proposed tool with existing tools is a quite cumbersome and difficult task, since a lot of publications can be found on this subject, some of them dated more than ten years ago. An overview of three "early-birds" is given in [Carl Spec88], where IDAC ([Degr JSSC87]), OPASYN ([Koh TCAD90]) and OASYS ([Harj TCAD89]) are compared. All based on a fixed library of unsized schematics, they focused on speeding up the design of commonly used analog blocks. As is also the case for the knowledge-based BLADES-tool ([El-T TCAD89]), none of them supported the user *during* the design of circuits of which the schematic and/or expert knowledge was not entered in the library yet. A tool that was focussed more for use with new schematics was DELIGHT.SPICE [Nye TCAD88]. Being an optimization-based system, it still *requires the formulation of the design problem as a certain standard mathematical programming problem* (sic). As is also the case for tools as ASTRX/OBLX [Ocho TCAD96] and methods as [Gielen JSSC90] using simulated annealing as optimization engine, the input of expert knowledge concerning circuit design is not foreseen. The same holds for more exotic types of synthesis tools as FASY [Torr TCAD96].

In *recent publications* a stronger polarization can be observed between approaches that go for fast results on the one hand and approaches that try to design the circuit using the same simulation environment as used for circuit validation on the other hand. The already mentioned method of geometric programming belongs to the group of automation approaches that opt for a fast result and has recently been used to optimize different types of circuits in a very short execution time ([Hers TCAD01, Hers DAC99, Hers ICCAD99]). As was also the case for DELIGHT.SPICE, for this approach a very specific mathematical formulation of the circuit has to be written for each new circuit topology. Moreover, not all circuits can be represented as a geometric program. The other group of approaches contains tools as MAELSTROM and Anaconda developed for the simulation-based synthesis of custom analog cells. Also in this group are tools as [Vanco DAC00] and [Vanco ICCAD01], the latter proposing a methodology that takes layout-related parasitics directly into account during the sizing process of the circuit. Still, none of these tools enable the designer to use his knowledge about the circuit to directly steer the optimization process.

In the environment proposed here, a combination is made between the equation or plan-based and the simulation or optimization-based solutions resulting in a method that can have the speed of plan-based methodologies without sacrificing accuracy. The knowledge of the designer is captured at the level a designer is working at. There is no need for simplified analytical expressions of the circuit behavior as in the exclusively plan-based tools, and due to the inclusion of expert knowledge the number of invalid circuits being generated is reduced as compared with the exclusively optimization-based solutions. The use of a Design Template also enables the inclusion of layout-related issues during the "execution" of the template, being the circuit sizing step.

The direct integration of the tool within the everyday design environment of an analog designer is a quality found also in *commercial products* being promoted recently. These encompass:

- Specification-driven synthesis tools based on the use of an extensive library [Barcelona] or on the use of "synthesis plans" [Antrim];

- Tools that directly implement the everyday design flow of an analog designer and that interact with the design and simulation environment the designer is using [ADA, NeoLin]. The design flow of these tools in fact shows great similarity with the methodology proposed here;

- Tools somewhere in between the two types above: starting with a schematic entry within the "normal" design environment, but then using proprietary methods for the next design steps [ComCad].

However, none of these tools is based on the use of a Design Template that is specific for a certain topology. Therefore, the same drawbacks as compared to the tools found in recent literature still hold: no topology-specific knowledge can be captured in an "executable" form.

A final conclusion of this historical perspective is that the introduction of a Design Template consisting of a circuit schematic and a set of Design Directives that capture the design strategy of the expert designing the circuit, has advantages over methodologies found both in literature and in recent commercial tools. The usability of the proposed method is demonstrated in the next section for a CMOS LC-type VCO circuit topology.

3.4 The CYCLONE Tool

3.4.1 Introduction

The previous section described the DLE methodology and proposed the implementation of an environment that can be used to help the designer in his effort of optimizing newly developed circuit topologies. This section gives a detailed example of a tool that uses the elements of the environment as described above. Two design templates for the CMOS LC-type VCO's shown in Fig. 3.10 have been written and used within an optimization loop to obtain a VCO with a certain phase noise and tuning range while minimizing the power consumption.

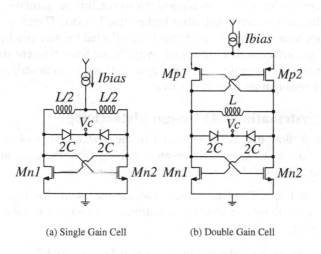

(a) Single Gain Cell   (b) Double Gain Cell

Figure 3.10. VCO templates included in CYCLONE

In this tool, all layout parasitics of the core devices of the oscillator are taken into account. The design template indicates how the gain cell and varactor diode of the oscillator have to be sized, given the inductance and parasitic resistance of the coil being used. The only independent parameters remaining are the geometrical parameters of the coil: width, number of turns, radius and possibly also the number of metal layers used in parallel. Using a predecessor of the tool presented here ([DeMuer ICECS99]), already state-of-the-art VCO's have been designed and reported ([DeMuer ESSCIRC99]).

After some words on the used method of inductor parameter prediction in a first subsection, the second subsection explains the systematic design methodology for VCO's used by the tool. The third subsection explains in detail the sizing procedure and the cost function which are implemented in the design template as expert knowledge together with the systematic design methodology. The next subsection deals with the layout generation consisting of a generic part and a knowledge-based part that belongs also to the design template. Before the conclusions, a subsection with examples and measurement results is included.

3.4.2   Inductor Parameter Prediction

Since any simulation can only be as accurate as the models used for the devices that are simulated, a good inductor parameter prediction is important for LC-type VCO's. The optimization process used by the tool needs 500 to 800 circuit evaluations to stabilize, and therefore the simulation time of a single circuit should be kept as low as possible.

For the reasons already mentioned in Section 2.3.3.3 FastHenry [FastHenry] has been selected for the coil simulation. However, this program only gives reliable results for substrate resistivities higher than 5 Ω.cm. Therefore, the finite element simulator Magnet [Free 93] has been selected for parameter prediction of coils on substrates with a lower resistivity. Since finite element simulations can be very time consuming, a special way of coil geometry description is used to reduce the simulation time in this case.

3.4.3   Systematic VCO Design Methodology

The design flow that is implemented in the CYCLONE tool is shown in Fig. 3.11. This flow has been automated, starting from the user inputs. The Optimization Start-up phase needs the following inputs:

- The wanted Specifications. These include the oscillation frequency, the tuning range, the phase noise specification and the supply voltage, as shown in Table 3.1;

- The used technology, defined in its *Custom Technology File*;

- The preferred oscillator topology (see example topologies in Fig. 3.10).

For each new technology, a *Custom Technology File* has to be generated. Physical and electrical properties of the used technology, e.g. metal, substrate, oxide thickness and sheet resistances, have to be provided by the user and are recalculated automatically to the technology parameters needed by the tool.

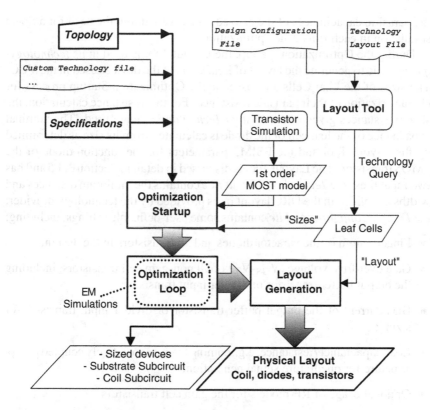

Figure 3.11. Flowchart of the CYCLONE tool

Table 3.1. Typical specifications for a VCO for RF-applications

Oscillation Frequency (F_c)	2 GHz
Phase Noise @$Fc + 600$ kHz	< -120 dBc/Hz
Tuning range	20%
Power Supply	2 V
Power consumption	Minimal

Based on the selected topology and technology, a first-order model for the current, G_M and capacitances of the gain cell is derived from a single Spice simulation per gain cell transistor. The first-order model is a linearization in the DC-operating point, that expresses current, G_M and capacitances as a linear function of the transistor width. In the Optimization Loop, this first-order model is used to calculate the needed size of the gain cell transistors and to determine the parasitic capacitances of the resized transistor without the need for a Spice simulation in each run of the optimization.

During the Optimization Startup, the Layout Tool is used in its *technology-query* mode to generate the layout of Leaf Cells of the devices used in the VCO. The sizes of the Leaf Cells are passed to the Optimization Startup process for characterization of their parasitic resistance. For this resistance calculation, the sheet resistances given in the *Custom Technology File* are used. The nominal capacitance of the leaf cell of the diode is calculated from its size as determined by the Layout Tool and the BSIM3 parameters for the junction diode of the pMOS transistor. The Layout Tool is discussed in detail in Section 3.4.5 and has two input files: the *Technology Layout File* contains the minimum distances and widths as defined in the DRC layout rules provided by the technology provider; the *Design Configuration File* contains some user-definable settings, including:

- Finger length of the varactor diodes and all transistors in the design;

- Gate overdrive voltage (V_{GS}-V_T) of all non-gain cell transistors, including the output buffer transistor or divider input transistor;

- Bias current of the output buffer transistor or divider input transistor for sizing;

- Load capacitance/resistance, e.g. coming from a capacitively coupled polyphase structure for quadrature signal generation;

- Optional usage of RF-models for the gain cell transistors.

In fact, the *technology-query* mode of the Layout Tool replaces the manual layout of the basic leaf cells and the parasitics calculation of these cells normally performed by hand by the designer once for each design and each new technology. For a single design the gain in time is rather limited, but it enables the use of the tool as a VCO-block generator within a high-level analog design and synthesis framework. The same method to incorporate layout already during design is used in [Vanco ICCAD01], but there the layout generation step has to be repeated in each design evaluation step due to the absence of a leaf-cell based design.

Since the output buffer of the VCO or the input transistor of an optional divider following the VCO is sized during the Optimization Startup its exact parasitic capacitance can be added to the LC-tank.

After the Optimization Startup the Optimization Loop which is described in detail in the next section, is started to determine the optimal device sizes. In each run of this loop, the EM simulator is used to calculate the electrical parameters of the coil. The output of the Optimization Loop is a fully sized VCO circuit, a lumped-element subcircuit of the coil as shown in Fig. 2.16 of Section 2.3.3.5 and a substrate subcircuit.

In the last phase, the Layout Tool in *layout-generation* mode composes the different devices, starting from the previously generated leaf cells. After a placement step based on a template suited for the chosen VCO topology, and after routing, the layout tool returns as final result the GDSII file of the VCO circuit.

The main program of the CYCLONE tool is programmed in TCL/TK ([TCL]), a platform-independent scripting language. The advantage of using this scripting language is the rapid development of a program involving a graphical user interface[6]. Since the TCL/TK script used by CYCLONE does not perform extensive calculations or time-critical tasks, the slower execution time inherent to an interpretation-based scripting language does not significally slow down the program execution. A dedicated C-program has been written for the generation of the Magnet and FastHenry input files. The code for the layout generation has been written in C++ as an extension of the in-house layout environment [Lamp PhD99].

3.4.4   Optimal Sizing of the complete VCO circuit

3.4.4.1   Overview of the Circuit Optimization Loop

Fig. 3.12 shows the small-signal schematic of a basic oscillator with an LC-tank. This representation is valid for both topologies of Fig. 3.10 and can be regarded as the small-signal schematic of a generic LC-oscillator. Five parameters fully characterize its behavior: Ls, Rs, Cp, Rp and G_M. In the optimization loop of CYCLONE, these parameters are *dependent parameters* that are calculated from the *independent parameters* and a set of *constants* in each run of the optimization loop. Table 3.2 gives an overview of the parameters involved in the optimization process. It should be noted that the CYCLONE tool uses a first-order MOST model to reduce evaluation time. These first-order calculations could be replaced by a more accurate device sizing step, but the accuracy of the implemented method proved to be sufficient. As indicated in the last line of Table 3.2, the Cost Function is calculated using the Power, Tuning and Phase Noise performance parameters of the oscillator. If needed, other performance parameters as area, oscillation amplitude, etc. could be added easily.

The *independent parameters* of the optimization loop are the topological parameters of the coil: *radius, metal width, number of turns* and the *number and type of used metal layers*. They are clearly indicated in Fig. 3.13 that shows the interaction between the different calculation steps in the VCO optimization

---

[6]The popularity of TCL/TK to write proprietary extensions to existing design frameworks can be deduced from the fact that recently courses are being organized ([ECD 02])

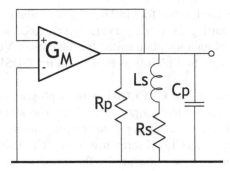

Figure 3.12. Simplified circuit of a generic LC-oscillator

Table 3.2. Summary of the parameters used in the optimization process

User-Definable Constants	<ul><li>*Safety*: The Gain Margin of the oscillator to ensure start-up</li><li>*Vgs-Vt, L, Finger Length*: MOST parameters of the gain cell</li><li>Min. *Tuning*, Max. *Phase Noise*: Oscillator Specifications</li><li>*Vdd*: Power Supply Voltage</li><li>*Swing*: Estimated oscillation amplitude</li><li>Technology Parameters: Layer thicknesses and electrical parameters</li></ul>
Derived Constants	<ul><li>G_M, Cap. and Ids per unit width of the gain-cell transistors. *This is a 1st-order MOST model, used in the optimization*</li><li>Leaf Cell Parameters: Rs of gain and diode Leaf Cell Capacitance of diode Leaf Cell</li></ul>
Independent Parameters	<ul><li>*Width, Number of Turns, Radius*: Topological coil parameters</li><li>*Number and Type of used metal layers* of the technology used for the coil</li></ul>
Dependent Parameters	<ul><li>*Ls, Rs, Cp, Rp*: Electrical coil parameters</li><li>*Rdiode, Number of Diode Leaf Cells*: Diode parameters</li><li>*Rgate, Number of Gain Leaf Cells*: MOST gain-cell parameters</li><li>G_M, *Power, Tuning, Phase Noise*: Oscillator performance parameters</li></ul>
Cost Function	<ul><li>Is calculated using the *Power, Tuning* and *Phase Noise* Parameters</li></ul>

loop. A double arrow is used to represent a recalculation of parameters from the output format of the previous block to the input format of the next block.

The optimization loop is initialized using either a randomly generated or a user-defined Start Coil, characterized by its topological parameters. An opti-

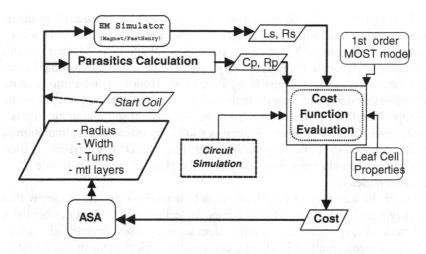

Figure 3.13. The VCO circuit optimization loop

mizer is used to search for a global minimum in the cost function. In our case, the Adaptive Simulated Annealing package ([ASA]) is utilized as optimizer core, because of its ability to find a global minimum in the presence of multiple local minima in a multi-variable system. The topological parameters are converted to electrical parameters (*Ls, Rs, Cp, Rp*) that are used for the calculation of the cost function. In the cost function evaluation the leaf cells are scaled according to the VCO-design rules implemented in the CYCLONE tool. This scaling is based on the first-order MOST model and the leaf-cell parameters calculated during the Optimization Setup phase. Therefore, no Spice simulation or additional runs of the Layout Tool are needed inside the Optimization Loop. However, optionally circuit simulation can be performed in each run of the Optimization Loop to come to a more accurate cost evaluation, at the expense of a longer execution time. The result of the cost function evaluation is sent back to ASA that generates the next set of values for the independent parameters.

3.4.4.2 Conversion of Coil Topological to Electrical Parameters

The conversion of the *independent* coil topological parameters into the *dependent* electrical parameters *Ls, Rs, Rp* and *Cp* is done using two parallel processes. *Ls* and *Rs* are the coil inductance and its total series resistance; *Cp* is the combined parasitic capacitance of the coil and substrate, and *Rp* is the parallel resistance to ground. Of the two parallel processes, the first one interfaces with the EM simulator, the output of which is converted to the parameters *Ls* and *Rs*. The second process calculates the parasitic elements of the coil using analytical formulae.

In CYCLONE, two different EM simulator tools are integrated to determine the electrical coil parameters Ls and Rs. For technologies with a low-resistivity substrate (< 5 Ω.cm), the finite-element simulator Magnet is used; for technologies with a high-resistivity substrate the three-dimensional inductance extraction program FastHenry is utilized. Both tools take into account all substrate effects and eddy currents in the metal lines, yielding very accurate results. FastHenry computes the frequency-dependent self and mutual inductance and resistance between conductors and is optimized for the simulation of planar conductors on planar ground planes. In CYCLONE, FastHenry is used to simulate the inductor as it actually is on the physical layout, including all its connecting leads.

Since the accuracy of FastHenry is not ensured for ground planes with a resistivity $\rho_{sub} < 5$ Ω.cm, Magnet is used for technologies with a low-resistivity substrate. This program uses a finite-element method to calculate inductances and resistances, resulting in a longer computation time than the method used by FastHenry. To speed up the simulation in this case, the axi-symmetric properties of circular coils are exploited. Only a two-dimensional simulation of a cross-section has to be performed to obtain full three-dimensional information [Cran CICC97]. The disadvantage of this method is that only the central part of the inductor is simulated accurately with Magnet. For the leads connecting the inductor center with the gain cell and the varactor diode, less accurate analytical formulae are used. Very reliable results can still be obtained for frequencies in the range of 1 to 3 GHz, because fairly large inductors are needed there, with a relatively big inductor center in comparison with the connecting leads, making the losses in the central part dominant over the losses in the connections. Therefore, this implementation is a good compromise between speed and accuracy.

For both low- and high-resistivity substrates, an inductor simulation and parameter extraction takes about 2 minutes of simulation time on a HP 9000/785 machine.

The coil model used inside the optimization loop is shown in Fig. 3.14. Herein, Cox is the physical coil-to-substrate capacitance, $Rsub$ and $Csub$ are the substrate resistance and the substrate parallel capacitance. This three-element parasitic model is converted to the two parallel elements Rp and Cp of Fig. 3.12 by recalculation at the oscillation frequency.

The model of [Meijs Int84] for micro-strip lines is used for the calculation of the parallel coil-to-substrate capacitance per unit length:

$$Cox \approx \varepsilon_{ox} \cdot \left[\frac{W}{H} + 0.77 + 1.06 \left(\left(\frac{W}{H} \right)^{0.25} + \left(\frac{T}{H} \right)^{0.5} \right) \right] \qquad (3.3)$$

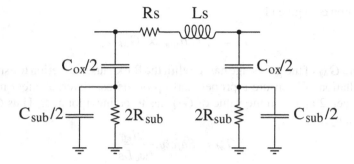

Figure 3.14. Lumped coil model

Herein, W, H and T are the width, the height above the substrate and the thickness of the metal strip (possibly consisting of a parallel connection of multiple metal layers), respectively, and ε_{ox} is the permittivity of the oxide.

3.4.4.3   Cost Function

The cost function that has to be minimized is made up out of three elements: power, phase noise and tuning range. It is given by:

$$Cost = PRC * Power + NRC * NC + TRC * TC \qquad (3.4)$$

with:

- *PRC, NRC, TRC*: relative weighting coefficients for the power, phase noise cost and tuning cost respectively;

- $NC = 1 + abs(noise - noisespec) * \frac{noise - noisespec}{noisespec^2}$ (phase noise cost);

- $TC = 1 + abs(tuning - tuningspec) * \frac{tuning - tuningspec}{tuningspec^2}$ (tuning cost).

The relative weighting coefficients for both noise and tuning are non-symmetrical, i.e. a low value is used when the specification is met and a high value when it's not (as kind of penalty function). The calculation of the three elements in the cost function is described here:

- Phase noise and power depend directly on the value of R_{eff}. This effective tank resistance consists of two terms:

$$R_{eff} = Rs + \Delta Rs \quad \text{where} \quad \Delta Rs = Rdiodes + Rgate \qquad (3.5)$$

is the additional resistance in the LC-tank caused by the implemented diodes and the gates of the gain transistors. The value of *Rgate* that has to be taken into account is only 1/3 of the physical gate resistance due to the distributed nature of the gate ([Tsiv 99]).

■ The power is given by:

$$P = V_{dd} \cdot I_{bias} \propto G_M \cdot V_{dd} \qquad (3.6)$$

where G_M of the gain stage has to fulfill the Barkhausen criterion to establish oscillation. To ensure a proper start-up of the oscillator, a safety margin between 2 and 3 on the value of G_M has to be incorporated. Thus G_M is given by:

$$G_M = Safety \cdot \frac{R_{eff}}{(\omega_o Ls)^2} \qquad (3.7)$$

with ω_0 the oscillation frequency. In each cost function evaluation, R_{eff} is calculated in an iterative process based on the first-order MOST model. G_M is calculated using equation ((3.7)) replacing R_{eff} by:

$$R_{eff}^{(i)} = Rs^{(i)} + \Delta Rs^{(i-1)} \quad \text{with} \quad \Delta Rs^{(0)} = 0, \qquad (3.8)$$

ΔRs is initialized at zero for the first calculation of G_M. Then, the following steps in the calculation of ΔRs (or R_{eff}) can be distinguished:

–   The needed width of the gain transistors is calculated from G_M using the first-order MOST model generated in the optimization loop start-up phase;

–   Next, the current and the parasitic capacitance of the gain transistors are calculated, using the same MOST model. The gate resistance is calculated from the number of leaf cells needed to implement the gain transistors and from the resistance of one leaf cell;

–   Then, the total parasitic capacitance is calculated. From this, the needed additional diode capacitance C_{diode} to obtain the wanted oscillation frequency is deduced;

–   Finally, the diode resistance is calculated from the number of diode leaf cells needed to realize this capacitance.

The previous steps are repeated, starting with the calculation of G_M using the newly calculated R_{eff}, until the relative error on G_M is smaller than a certain preset value. Since these calculations are fully equation-based, no additional Spice simulations are needed to obtain the final value of R_{eff}.

■ The second component in the cost function, the phase noise, is given by [Cran TCAS95]:

$$dV_{out}^2\{\Delta\omega\} = kT \cdot R_{eff} \cdot [1+F] \cdot \left(\frac{\omega_0}{\Delta\omega}\right)^2 \cdot \left(\frac{1}{swing^2}/2\right) \qquad (3.9)$$

where $dV_{out}^2\{\Delta\omega\}$ is the single-sideband spectral noise density at an offset Δ from the oscillation frequency ω, F is the excess noise of the oscillator's amplifier, *swing* is the differential oscillation amplitude and R_{eff} is the LC-tank's effective resistance. Only the $1/f^2$ – region of phase noise is taken into account by this formula; 1/f-noise upconversion is not calculated in the current version of the CYCLONE tool. The latter is only required for applications with a specification on the phase noise very close to the carrier. For the phase noise calculation, an estimate of the oscillation amplitude is needed. Normally, a user-supplied estimate of the *swing* value is used. Optionally, one transient and one AC Spice simulation can be performed automatically during the cost function evaluation to have a more accurate value for the oscillation amplitude and the effective resistance R_{eff}.

- The last element of the cost function, the tuning range, is calculated as:

$$Tuning = \frac{\Delta\omega}{\omega_c} \qquad (3.10)$$

with

$$\Delta\omega = \omega_0 - \omega_{min} \quad \text{and} \quad \omega_c = (\omega_0 + \omega_{min})/2 \qquad (3.11)$$

The value of ω_{min} is calculated as:

$$\omega_{min} = \sqrt{\frac{1}{Ls/2(C_{0,diode} + C_{par})}} \qquad (3.12)$$

Herein, C_{par} is the total parasitic capacitance of the oscillator comprising the gain cell capacitance and the capacitance of any (buffer) circuits connected to the outputs of the oscillator. $C_{0,diode}$ is the capacitance value of the varactor diode under a zero DC voltage bias. Using the pMOST junction diode model, it is calculated from the (minimum) diode capacitance C_{diode} needed to obtain the wanted maximum oscillation frequency ω_0.

Obviously, this detailed calculation in first order reduces to:

$$Tuning \propto \sqrt{C_{diode}/C_{tot}} \qquad (3.13)$$

which can be recalculated to directly show the inductance value Ls and the wanted oscillation frequency ω_0:

$$Tuning \propto \sqrt{C_{diode} \cdot (\omega_0^2 \cdot Ls/2)} \qquad (3.14)$$

3.4.4.4 Evaluation Time

Due to the leaf-cell-based sizing method and the parameterization of these leaf cells during the Optimization Setup, the consecutive evaluations of the cost function within the Optimization Loop do not require any circuit (Spice) simulations in normal operation. The optional Spice simulations that can be performed at each cost function evaluation have been found not to have enough added value to justify the increase in evaluation time. Therefore, the simulation time of the coil is dominant on the whole cost function evaluation, resulting in a total evaluation time of 1 to 2 minutes for a single run of the Optimization Loop on a HP 9000/785 machine. Since a simulated annealing process takes about 500-700 evaluations to come to a stable solution, about 10-24 hours are needed for a complete VCO optimization. The use of Matlab as simulator for an inductor, using a proprietary set of functions, leads to a significant reduction of evaluation time. In less than 12 hours, more than 4000 evaluations can be accomplished with such a setup. The accuracy of the final result then solely depends on the reliability of the models incorporated within the set of Matlab functions.

As a comparison, an indicative time is given for a formula-based optimization approach using a compiled (C/C++) program instead of a script-based program. This time is estimated to total a few minutes only. It clearly indicates that the use of a good formula-based model for the symmetrical octagonal coil of Fig. 2.10 (Section 2.3.3.3) would enable the use of the CYCLONE tool as a kind of power-area-performance estimator within an analog high-level design environment.

3.4.5   Layout Generation of the Complete VCO Circuit

In the previous section the complete VCO circuit has been sized in an automatic way, which reduces the design time considerably. We will now discuss how the layout is generated automatically. Note that an automated layout estimation capability (i.e. running the layout generator in query mode) is already used during the sizing to take actual layout parasitics into account.

In most analog circuit layout tools the following procedure has been adopted: module generation, module placement and routing (optionally followed by analog compaction) [Chang PhD95, Gielen Proc00, Gielen PhD91]. Procedural module and device generators are widely accepted in industrial environments [Pcells]. Optimization-based analog place and route tools [Lamp PhD99, Cohn JSSC91] have also been proposed in academia. These tools are directly or indirectly steered by analog constraints: symmetry in balanced circuits is enforced at layout level (both in placement and routing), the parasitic capacitance and resistance of nets is bounded and wire sizing is adapted to the currents carried. These tools are used as baseline to automate the layout generation of RF (VCO) circuits. From the schematics (see Fig. 3.10) it could be assumed that a VCO circuit contains standard (analog) CMOS devices, except for the integrated inductors. In reality only the current source is implemented as a standard (non-RF) device; the other devices are grouped in RF-optimized layout structures. E.g. specific layout structures for a balanced diode pair and gain cell are described in Section 4.2.2. For the designed VCO's this results in 4 out of 5 layout structures being optimized for RF and thus requiring specialized procedural device generators.

Since RF applications benefit strongly from CMOS deep submicron technology improvements (high f_T, lower parasitics), the designs are prime candidates for migration towards the newest state-of-the-art technology. Technology independence is thus a must for RF layout synthesis if it is to be automated. In this respect it must also be noted that in the rapidly evolving RF world, new layout structures are introduced frequently. Therefore, in CYCLONE a module generator technology is used that combines technology independence with high flexibility to introduce new types of layout structures. These two features differentiate the approach from other widely used pcell approaches [Pcells]. This will be described in more detail further on.

3.4.5.1 Module Generator Technology

The three module generators required for the layout automation of CMOS LC-tank VCO's have been implemented in our in-house layout framework [Lamp PhD99]. The coil device generator is designed to generate fully balanced coil geometries as shown in Fig. 2.10 of Section 2.3.3.3 and accepts as input the geometrical coil parameters: the coil radius, the wire width, the number of turns and the used metal layers. It supports an unlimited number of metal layers in parallel. The device generator reads the DRC (*Design Rule Check*) rules from the *Technology Layout File* (Fig. 3.11). This guarantees technology independence. The generator has been implemented in C++ and consists of about 1000 lines of code in the layout framework [Lamp PhD99].

Whereas the coil is a concentric device, the varactor and gain cell are block repetitive. This block-repetitive property has been exploited in the implementation of a new type of device generator: a template-based procedural generator that employs block stretching. This will be discussed next.

3.4.5.2 Template-Based Block-Repetitive Module Generator

Three components have been used to create this new type of device generator:

1 The stretch box and command. In Fig. 3.15(a), an example of a very basic hypothetical device template is shown, consisting of some rectangles in full lines. Within this template, two stretch boxes are defined, shown in dashed lines. A stretched layout can simply be generated with commands. A command *moves*, *cuts* or *copies* the area enclosed by the selected stretch box. This makes it a much more powerful operation than the (interactive) stretch operations implemented in [Oust DAC84] and [Pcells]. In Fig. 3.15(b), the result of a double, downwards copy operation executed on the selected

(bold) stretch box of Fig. 3.15(a) is shown. In the associated command file, one line of code is needed for this operation, given in Fig. 3.15(c). If the stretch box has a height equal to the manufacturing grid of the technology, this combined operation is equivalent to stretching of the geometry;

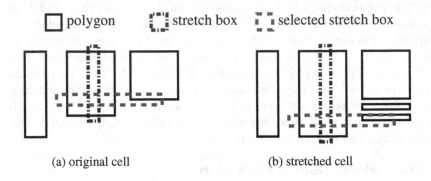

(a) original cell   (b) stretched cell

CP select stretchbox1 direction=down times=2

(c) One line of code in associated command-file

Figure 3.15.   Example of a stretch operation on a cell

2 Symbolic layers ([Oust DAC84]). These are needed to ensure technology independence. Otherwise, a new template for every device would be required each time the layer names change, e.g. when switching technologies as is the case in [Pcells]. A symbolic layer is for instance a contact area between gate polysilicon and metal1. When a box is drawn on the symbolic layer of the device template, it is replaced in the generated layout by a structure containing polysilicon, metal1 and a number of contacts. The replacement of a symbolic layer with the layers of a specific technology is done taking into account the DRC rules, as defined in the *Technology Layout File* (Fig. 3.11);

3 A technology-independent device template. Globally, two distinctive steps can be distinguished in the layout generation of a custom device. In Fig. 3.16(a) the original device template for the gain cell, drawn using symbolic layers, is shown. A stretch box is indicated in dashed line. The first step is the *technology-query* mode, in which the technology-specific leaf cell of Fig. 3.16(b) is generated. This mode is used during optimization start-up to extract minimal dimensions for the gain cell and the varactor diode, as has been explained in Section 3.4.3. The result of the second step in the layout

generation (the actual *layout-generation* mode) is a complete automatically generated gain cell, as shown in Fig. 3.16(c). This step incorporates a re-sizing of the leaf cell to the dimensions imposed by the CYCLONE tool, followed by a procedural copy and placement step. Indicated with the rect-angle in the lower right corner of Fig. 3.16(c) is the resized leaf cell of Fig. 3.16(b) while the full cell is shown in Fig. 3.16(c). All operations needed for this resizing are implemented using stretch boxes.

The template-based module generation program has been implemented in C++ code and is 3000 lines long [Lamp PhD99]. The implementation of the gain cell generator has then been realized in less than 50 lines of code, similar to that of Fig. 3.15(c). The execution times are less than 1 second on a PIII/500 MHz Linux PC. The same principle has been used to implement the varactor diode structure. Since it is less complex, its associated code is only 30 lines long. Both gain cell and varactor diode generator are considered as a part of the Design Template, since they contain detailed knowledge about the layout generation step of these specific components of the VCO. However, these parts of the Design Template differ from the rest since they can be reused in com-pletely different circuits, like RF buffers, whereas the calculation of the gain cell properties are really specific for the Design Template of an oscillator.

3.4.5.3   Procedural VCO Placement & Route

Based on a specific floorplan for each of the two topologies of Fig. 3.10, the placement and routing of the VCO circuit is done with the LAYLA tool [Lamp PhD99]. The relative positions of the modules in the floorplan are enforced during placement by extra constraints. A full layout generated for the single-differential circuit of Fig. 3.10(a) is shown in Fig. 3.17. This layout in GDSII format is the output of the CYCLONE tool.

3.4.5.4   Total VCO-Design Time

It takes only 5 seconds of CPU time to generate the final layout on a PIII/500 Linux PC. Obviously, this time added to the time of about 2 minutes needed for the Optimization Setup is negligible compared to the time needed for the optimiza-tion loop to finish. The time to design and layout an optimized VCO thus totals less than 24 hours. An experienced designer needs about three days to arrive at a comparable result, assuming he has some layout tools available to generate the symmetrical octagonal coil, the custom gain cell and the balanced diode layout. If only the standard layout of a MOS transistor can be generated automatically, the design and layout time almost doubles to arrive at the same result.

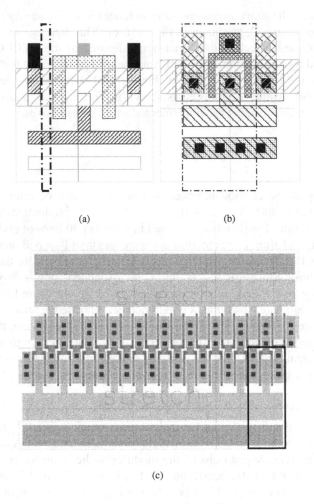

(a)   (b)

(c)

Figure 3.16. Device layout template before (a) and after (b) technology query, and (c) final layout of a cross-coupled pair

3.4.6   Design Examples - Experimental Results

To demonstrate the use of the tool, two design experiments are given. The first example gives a comparison at the simulation level between a technology with high- and one with low-resistivity substrate. This experiment has already been described briefly in Section 2.2.3 to demonstrate the positive influence of a high-resistivity substrate on the performance of analog RF circuits. In a second example, a comparison of measurement results and simulations of a VCO designed using CYCLONE, is shown.

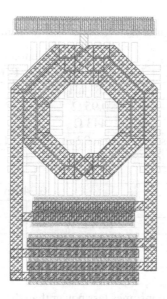

Figure 3.17. Automatically generated layout of the VCO core circuit

Table 3.3. Technology features

Technology	CMOS 0.25μm	
	"Test" version	"Normal" version
Substrate resistivity	0.0143 Ω.cm	14.3 Ω.cm
Available metal layers	4 layers *(used layers = design variable)*	

As a first example, the usability of the tool for comparison of different technologies is demonstrated. A VCO design is performed in both a technology with high-resistivity substrate and one with low-resistivity substrate. To clearly show the substrate influence on the optimal design, the resistivity of the substrate is changed in an existing technology keeping all other technology parameters the same. This kind of experiment can be useful for customizing or choosing a technology for a certain application. In Table 3.3 the main technology features are given. A VCO with the design specifications of Table 3.1 in Section 3.4.3 is generated using CYCLONE. The leads are assumed to be 90 μm, and the used topology is the one given in Fig. 3.10(a), using only nMOS transistors for the gain cell.

After 700 evaluations, corresponding to a total simulation and optimization time of about 24 hours on a HP 9000/785, the results shown in Table 3.4 have been obtained. The inner radius defines the size of the open area inside the inductor turns. From this table, it is clear that the VCO in the technology with a

Table 3.4. Optimization results of high- vs. low-resistivity substrate

Parameter	Technology	
	Low-resistivity sub	**High-resistivity sub**
Ls	1.81 nH	2.85 nH
Rs	0.95 Ω	0.74 Ω
Rpeff	443 Ω	309 Ω
Inner Rad, W, Turns	134μm, 22μm, 2	178μm, 18μm, 2
$gm_{transistor}$	5.6 mS	8.1 mS
$W_{transistor}$	46.7μm	66.8μm
$L_{transistor}$	0.24μm	
Used metal layers	3 top layers	all 4 layers
Power	12.8 mW	8.8 mW
Phase Noise (calc.)	-120.9 dBc/Hz	-120.1 dBc/Hz
Tuning (calc.)	15.5%	15.6%

high-resistivity substrate consumes less power than the VCO with similar specifications in the low-resistivity substrate technology, due to its lower substrate losses[7]. From this, one can conclude that a high-resistivity substrate is certainly favorable for RF design as already stated in Section 2.2. The design with low-resistivity substrate also suffers more from (capacitive) substrate coupling, and this leads to an optimal inductor using only the top three metal layers, as opposed to the usage of all four metal layers in the case of the high-resistivity substrate. As shown in Table 3.2, the number of metal layers to be used is an *independent* design variable, that is determined during the optimization. CYCLONE thus automatically implements the remarks made in [Long JSSC97] concerning the use of the top-level metal layers to reduce substrate coupling in technologies with a low-resistivity substrate.

To demonstrate the progress of the optimization, a plot of the cost function, width, radius and turns of the optimal coil device versus the number of evaluations is shown in Fig. 3.18. Here also, the plotted radius is the inner radius of the coil.

As a second example, a VCO has been designed in a 0.35μm CMOS technology with a low substrate resistivity of 0.013 Ω.cm. The topology of Fig. 3.10(a) is used. In Table 3.5, the determined sizes and simulated specifications of the functional blocks of the optimized VCO are given. Note that in this case only metal layers 4 and 5 are used for implementing the coil.

---

[7]The small difference in phase noise can be explained by the lower noise contribution of the transitor gates due to a larger number of gain cell fingers, resulting in the higher power consumption

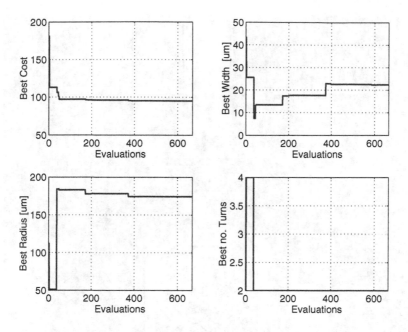

Figure 3.18. Progress of best inductor cost function, width, radius and turns as a function of the number of evaluations

Table 3.5. Sizes and specifications of optimal VCO-parts

Coil	**Radius (inner)**	53μm
	Turns	3
	Width	20μm
	Used layers	Metal 4 and 5
	Inductance	1.6 nH
	Series resistance	6.5 Ω
Diode	**W x L**	0.8μm x 71μm
	Fingers	20 per side
	Capacitance	3.5 pF
	Series Resistance	0.24 Ω
Gain cell	**W x L**	112μm x 0.35μm
	Gate Resistance	7.2 Ω

In Fig. 3.19, a chip photograph of the designed VCO is shown. The capacitors at the bottom are for decoupling of the power supply of the output buffer. In the center the inductor with three turns can be distinguished. On top of the inductor, the bias transistor is placed, and below the inductor are the diode cell, gain cell and output buffers.

Figure 3.19.   Chip photograph of VCO in 0.35μm CMOS technology

Table 3.6. Comparison of measurements and simulations

Technology	CMOS 0.35μm, low-resistivity substrate					
Vdd	1.8 V	**Ibias (core)**	11 mA			
Vc	**Oscillation Frequency**		**Error**	**Phase Noise**		**Error**
[V]	Simulation[a] [GHz]	Measured [GHz]	[%]	Calculated [dBc/Hz]	Measured [dBc/Hz]	[dB]
1.80	3.192	3.330	4.1	-115.4	-115.0	-0.4
1.69	3.161	3.301	4.2	-115.4	-114.0	-1.4
1.40	3.066	3.207	4.4	-115.4	-113.3	-2.1
1.14	2.967	3.101	4.3	-115.6	-111.1	-4.5
0.8	2.841	2.905	2.2	-114.5	-110.0	-4.5

[a]Simulated with Hspice

A comparison of the measurement results with the simulation results is shown in Table 3.6. The comparison is performed for identical *Vdd* and *Ibias* in both simulations and measurements. As can be seen in Table 3.6, the simulation results of the oscillation frequency are within 5% accurate compared to the measurement results. The calculation of the phase noise uses the R_{eff} of the circuit obtained from an AC simulation and the oscillation amplitude obtained from a transient simulation. The phase noise prediction is quite accurate for high values of *Vc*. Due to the EM analysis used here, these results are much better than the 40% error mentioned in [Cran PhD97] in the case an equation-based model is used. For lower values of *Vc* it is clear that a better varactor model is needed for an accurate simulation of their influence on the phase noise. A new model can be included seamlessly in the tool. A circuit simulation of the final design takes approximately 5 seconds for an AC analysis and 25 seconds for a transient analysis (simulated time of 15nsec) on an HP 9000/785.

For the design of VCO's at higher frequencies, the use of CYCLONE has been demonstrated in the successful design of a 17GHz oscillator in a 0.25μm CMOS technology [DeRa ISSCC01]. This design is discussed in Section 4.2 together with some specific RF-related design issues.

3.4.7 Conclusions

In this section a specification-driven, layout-aware design tool for CMOS RF LC-oscillators, called CYCLONE, has been presented. It synthesizes an optimal LC-tank oscillator design from specifications to layout and can be regarded as an example of the use of the template-based design optimization methodology presented in the previous section. The tool uses finite-element analysis to characterize the on-chip inductors more accurately in combination with the optimization power of simulated annealing. The layout parasitics are taken into account during the optimization of the circuit without the need for Spice simulations or layout generation steps in each run of the optimization loop, due to a parameterized leaf-cell-based design method. The layout is generated using a template-based approach with procedural and technology-independent module generators. The design experiments show the usability of the tool for a wide range of technologies.

Looking back at Fig. 3.4 introducing the DLE-methodology, it can be seen that from the interactions between block level, system and layout level all but one is implemented within CYCLONE. Interaction (3) from system level towards block level is the input supplied to the tool. The interaction with the layout level is completely automated, implementing interaction (4) as a layout generation step of leaf cells during the initialization phase of the optimization and interaction (2) as the construction of the sized VCO layout after the optimization and validation of the VCO is finished. The fourth interaction, being the feedback from block level to system level is only implemented what the area and power consumption is concerned. No high level model is generated by the tool. This is in full accordance with the fact that the template-based design methodology only implements the Block Sizing Function of the DLE Tool as defined in Section 3.2.3.

3.5   Final Conclusion

Starting from the vague and simple definition of what is considered in this work to be a functional block, a structured design methodology is proposed. It is called the DLE (Design-Level Entry) methodology since the point-of-entry of the methodology is situated at the level of the functional blocks an analog designer is working with, having a system level above it and a physical implementation level below. The main idea behind it is that a designer should be able to benefit from the use of the methodology without having to change his everyday design practice.

For the DLE tool an implementation of its Block Sizing Function is proposed that is based on the use of a Design Template consisting of a fixed topology and a set of Design Directives, being the knowledge of the designer about this specific topology of the functional block under consideration. Within the Design Template, layout information can also be included. Furthermore, the designer can indicate within the Design Template a limited set of independent parameters for which he wants an optimizer to determine the set of values that will minimize a cost function evaluating the circuit performance.

As an example of such a Design Template-based design, the CYCLONE tool for the automated design of optimized LC-tank VCO's is presented. The Design Template here consists of the VCO topology and the formula's needed to calculate the gain cell and varactor diode sizes given a certain set of inductor parameters. Layout parasitics are taken into account within the Design Template. The use of the tool has been demonstrated with some examples.

Chapter 4

VOLTAGE-CONTROLLED OSCILLATORS FOR HIGH DATA RATE APPLICATIONS

The presence of oscillators in communications systems is ubiquitous. Basically, the goal of a communication link is to exchange information between two separated points. This exchange of information is mostly done using some kind of modulated carrier signal. The modulating signal is a low power baseband signal that represents the information while the carrier signal with a distinct frequency is provided by an oscillator. After modulation, the carrier signal is further amplified until it has adequate power for transmission. Some systems, e.g. the Global Positioning System (GPS), use only a single carrier frequency for the information exchange. However, most systems use multiple carrier frequencies within a certain frequency band. To be able to transmit or receive on these different frequencies (called channels), multiple oscillation frequencies have to be available in the transmission and receive devices. To accomplish this without the need for multiple oscillators an oscillator with a variable frequency is used. Oscillators of which the oscillation frequency can be tuned by changing the voltage of a specific node of the oscillator circuit are called *Voltage-Controlled Oscillators* or VCO's.

In this chapter, VCO's are discussed that have been designed primarily for applications aimed at high-data rate transmission. In a first section an oscillator is presented for use in a broadband cable system designed to provide high-speed internet connectivity. Since the oscillator under consideration is used to drive a linear mixer block within a complete upconvertor and higher harmonics of the oscillator signal fall within the band of interest, the emphasis here lies on a low content of higher harmonics in the oscillator signal. The subject of the second section is the design of oscillators for high-frequency operation. These are needed in both high-data rate fixed-wireless applications as in high-speed optical telecommunications circuits. After a short description of typical problems arising at high frequencies, the design of a 17 GHz oscillator is presented. The use of the CYCLONE tool presented in the previous chapter showed very useful in the final design of this oscillator.

4.1   Oscillators for Broadband Systems

The definition of what has to be considered as a "broadband" system is a topic that has to be tackled before the design of an oscillator for such a system can be presented. In the first section the need for a sine wave oscillator signal with a high spectral purity in transmitter circuits for broadband systems, will be clarified. The second section focuses on the theory to design a linearized broadband oscillator, introducing the complex frequency domain description of a polyphase network and a quadrature square wave. The third section deals with a complementary technique to achieve a waveform with lower harmonic distortion components. In the last section a full CMOS realization on one chip implementing a linearized four stage ringoscillator optimized for high tuning range, is presented.

4.1.1   Broadband Systems

4.1.1.1   In-Band LO Harmonics

A system has to be treated as being a "broadband" system if the frequency band in which the signals of interest are being transmitted, is much larger than the bandwidth of the transmitted signal itself. This is clearly the case for a frequency range-expanded cable TV system as Hybrid Fiber Coax *(HFC)* that has an operational frequency band situated roughly between 100 MHz and 1 GHz and that works with a signal bandwidth of 5-6 MHz. The difficulty in the design of an oscillator for the transmission part of such a system is that the higher harmonics for LO frequencies in the lower range of the frequency band are still within the usable range, as illustrated in Fig. 4.1 for an *HFC*-like system.

In the design of the transmitter, care has to be taken to avoid the transmission of these unwanted signal components. When using a switching mixer, the output signal of the mixer will contain upconverted baseband signals at higher harmonics of the local oscillator signal, due to the switching behavior of the mixer cell.

A possible solution to get rid of the unwanted harmonics is to use a dedicated bandpass filter for each transmitted channel. These filters typically consist of high-quality discrete components, making them expensive. A solution that avoids the need for external filters could be the use of a non-switching mixer, e.g. a linear mixer [King JSSC97]. Here, care has to be taken in the generation of the LO signal, since the baseband signal is not only upconverted to the LO oscillation frequency, but also to each harmonic of the oscillation signal. Therefore, an LO signal with a high spectral purity is needed. This is explained in more detail in the next section.

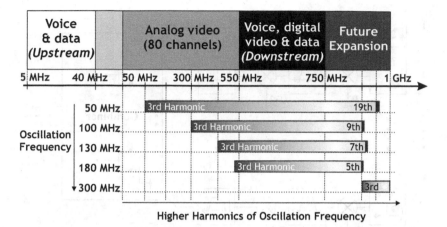

Figure 4.1. Frequency Chart of HFC-system with indication of in-band oscillator harmonics

4.1.1.2 Multi-Channel Transmission

Many different channels are available on a cable system as HFC. Traditionally, a cable TV system has been developed for use as a single-transmitter multiple-receiver system. The recent developments to use such a cable system in two directions, e.g. for high-speed Internet applications, does not fundamentally change this since the return path is situated in a small frequency band in the lower end of the available frequency spectrum. The broad frequency range going from 100 MHz up to 700-950 MHz is still used to transmit information in dedicated channels towards the consumer at home. These channels are injected in parallel on the cable in what is called a *Cable Head-End*. An example of a multi-channel upconversion scheme used in such a *Cable Head-End* is given in Fig. 4.2.

The information to be transmitted consists of video and data channels as visualized in the left part of the drawing. These are first upconverted to an intermediate frequency *(IF)*, bandpass filtered, upconverted to a channel-specific frequency and again bandpass filtered at this frequency. To end up with the multi-channel signal that is sent to the cable, a certain number of these channels is combined in an active or passive combiner. The use of a passive combiner results in a power loss that grows with the number of channels to be combined, because for each two channel combination, an insertion loss of 3dB has to be accounted for. Another drawback of this transmission scheme is the need for two bandpass filters per channel. To meet the stringent specifications on spurious in-band emissions and signal distortion, the second filter needs a very sharp transition band while maintaining a low distortion. Such a performance can

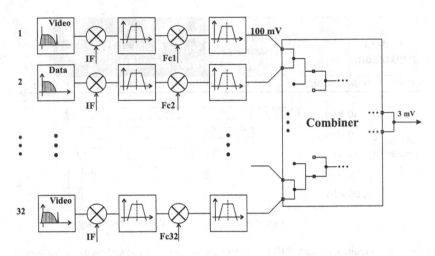

Figure 4.2.   Traditional head-end (multi-channel transmitter) of a cable system

only be reached using high quality discrete components, resulting in a high cost for each additional channel.

In the case that linear mixers are used, a more optimal transmission strategy with respect to a full integration in CMOS is possible. Since the linear mixers themselves do not generate any harmonics of the LO frequency, the bandpass filters (with discrete components) following the mixers in the topology of Fig. 4.2 can be omitted. Since a linear mixer is in first order a multiplier, the condition to have no upconversion components at the harmonics of the LO signal is that in the LO itself, these harmonics are low enough. This is mostly the case for a simple LC-type oscillator, but for other kinds of oscillators, e.g. a ringoscillator or an LC-type oscillator followed by a divider-type quadrature generator (Section 2.4), the harmonic content of the LO signal can be substantial. In that case, some filtering before or after the mixing operation, is still necessary. The possibility to perform the filtering on the quadrature LO signal, before the mixing operation, enables us to use a filter operation in the complex frequency domain. In this way, the properties of a quadrature signal in the complex frequency domain can be fully exploited to filter out unwanted harmonics. This complex operation will be explained in the next section.

4.1.2   Signal Linearization using a Polyphase Network

Using a multi-stage polyphase network, the *(4n-1)th* harmonics of a quadrature signal can be suppressed over a broad frequency range. This can be demonstrated by sending a quadrature square wave with a well-known spectral content through a polyphase structure. In Section 2.4.2 the use of a passive sequence-asymmetric polyphase network as a quadrature generator is described. There, a differential sinusoidal input source is applied to the network, and a quadrature signal is taken as output of the network. Thus, only two out of four input terminals are used. When a quadrature signal is already available, the functionality of the network can also be described as that of a complex bandpass filter. Such a quadrature input signal could be consisting of two differential signals with a 90 degrees phase shift, e.g. coming from a divide-by-two stage or from a ringoscillator. Here also, a quadrature signal is taken as output of the network. Thus, all eight terminals of the network are carrying an AC-signal.

To fully comprehend the operation of the polyphase network as a complex filter, first the definition of a complex signal and of a complex filter are given. Then, the analytical description of a complex square wave in the frequency domain is calculated. Next, the frequency behavior of the polyphase network as a complex filter is derived, followed by the explanation of the complex filtering that can be performed on a complex square wave.

4.1.2.1 Real and Complex Signals and Filters

Real Signals A good example of a *real signal* is the voltage on a node in an electrical circuit as a function of time. It has a certain amplitude and phase that vary in time. It is common knowledge that in the frequency domain this real signal has a frequency spectrum that is symmetrical around zero. When the real signal is designated by x_1 and its Fourier transformed by X_1, this is expressed analytically as:

$$X_1(j\omega) = X_1^*(-j\omega) \tag{4.1}$$

Complex Signals A *complex signal* is, simply put, a *pair* of independent real signals [Lee 1990, Bout RFD89]. In (4.2) the mathematical representation of a complex signal is given. Herein, x_1 and x_2 are real signals.

$$z(t) = x_1(t) + j \cdot x_2(t) \tag{4.2}$$

In the frequency domain, the complex signal can be written in a general form, given by:

$$Z(j\omega) = X_1(j\omega) + j \cdot X_2(j\omega) \tag{4.3}$$

In this case, the symmetry of the frequency spectrum is not axiomatically defined as it is for a real signal. On the contrary, it can be completely different for positive and negative values of ω. The Fourier expansion of the complex signal can be rewritten in such a way that a distinction between the frequency content for negative and for positive frequencies is made. For a periodic signal the description of the complex signal can be rewritten as:

$$z(t) = \sum_{n=1}^{\infty} (Z_n^{pos} * e^{jn\omega t} + Z_n^{neg} * e^{-jn\omega t}) \tag{4.4}$$

Quadrature Signals   As defined in Section 2.4.2, a *quadrature signal* consists of two signals that have a phase difference of exactly 90 degrees. Regarding them as two independent signals, they can be treated as one complex signal, using (4.2). In this text, arbitrarily we choose to let x_1 designate the signal that is *in*-phase and x_2 the signal that is in *quadrature*-phase when treating a quadrature signal as a complex signal. This choice has no influence whatsoever on the calculation results.

Complex Filters   A *complex filter* has a transfer function $H(j\omega)$ that is not necessarily symmetrical by definition. If the impulse response $h(t)$ of such a filter can be written as:

$$h(t) = h_{r1}(t) + j \cdot h_{r2}(t) \tag{4.5}$$

with both $h_{r1}(t)$ and $h_{r2}(t)$ real impulse responses, then the frequency domain description, or complex transfer function, of this complex filter can be expressed as:

$$H(j\omega) = H_{r1}(j\omega) + j \cdot H_{r2}(j\omega) \tag{4.6}$$

with H_{r1} and H_{r2} both real transfer functions, corresponding to the respective impulse responses. As is the case for the time description of a complex signal, the impulse response of a complex filter can be rewritten to clearly distinguish the frequency response for positive frequencies from the response for negative frequencies. In a general formulation the resulting Fourier expansion is given by:

$$h(t) = \int H_{pos}(f) * e^{jf\omega t} + H_{neg}(f) * e^{-jf\omega t} \, df \tag{4.7}$$

in which both H_{pos} and H_{neg} are complex functions of frequency.

Linear complex filter   In analogy with a linear operation in the frequency domain, the output of a linear complex operation can be described in the frequency domain as:

$$Y(j\omega) = H(j\omega) \cdot X(j\omega) \tag{4.8}$$

with $Y(j\omega)$ the complex output signal, $X(j\omega)$ the complex input signal and $H(j\omega)$ the complex transfer function. The following property of a linear system also holds for a linear complex filter:

$$f(c \cdot \alpha) = c \cdot f(\alpha) \tag{4.9}$$

with c a constant.

4.1.2.2   Complex Representation of a Quadrature Square Wave

As every quadrature signal, a complex square wave can be written as (4.2). As is chosen arbitrarily in Section 4.1.2.1, x_1 designates the square wave that is *in*-phase and x_2 the square wave that is in *quadrature*-phase, meaning that signal x_2 has a phase difference of 90^0 as compared to x_1. The Fourier expansion of the two real components can be written as (4.10) and (4.11):

$$x_1(t) = \frac{4}{\pi}(\cos \alpha t - \frac{1}{3}\cos 3\alpha t + \frac{1}{5}\cos 5\alpha t - \cdots) \qquad (4.10)$$

$$x_2(t) = \frac{4}{\pi}(\sin \alpha t + \frac{1}{3}\sin 3\alpha t + \frac{1}{5}\sin 5\alpha t + \cdots) \qquad (4.11)$$

Performing some complex calculus on (4.2) , (4.10) and (4.11) results in:

$$z(t) = \frac{4}{\pi}(e^{j\alpha t} - \frac{1}{3}e^{-j3\alpha t} + \frac{1}{5}e^{j5\alpha t} - \frac{1}{7}e^{-j7\alpha t} + \cdots) \qquad (4.12)$$

The third, seventh, ... i.e. *(4n-1)th* harmonics only appear with negative frequencies in the exponent, whereas the first, fifth, ... i.e. *(4n+1)th* harmonics only appear with positive frequencies in the exponent. In Fig. 4.3 the magnitude of the Fourier components of a quadrature square wave are visualized in the (complex) frequency domain.

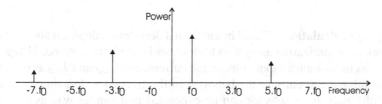

Figure 4.3. Complex Fourier spectrum of a quadrature square wave

The asymmetry in the frequency spectrum clearly demonstrates the complex nature of the quadrature square wave since a real signal always has a symmetric frequency spectrum. This property will be exploited by performing a complex operation on the signal. How this can be done using a polyphase network is pointed out next.

4.1.2.3   A Polyphase Network as a Complex filter

The Polyphase Network   In Fig. 4.4, a three stage polyphase network is shown. It has two real, differential inputs X_I, X_Q and two real, differential outputs Y_I, Y_Q. The behavior of this four terminal network as a complex filter can be described using (4.8), with:

$$X(j\omega) = X_I(j\omega) + j \cdot X_Q(j\omega) \qquad (4.13)$$

as a complex differential input signal and

$$Y(j\omega) = Y_I(j\omega) + j \cdot Y_Q(j\omega) \qquad (4.14)$$

as complex differential output signal.

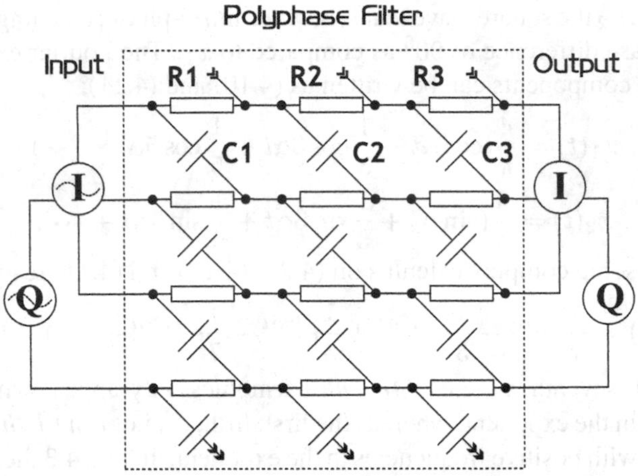

Figure 4.4.   Use of polyphase network as complex filter

Analytical calculation   To fill in the formal description depicted above, a full network characterization using Kirkhoff's laws has to be performed. Using V_{in} and I_{in} as the complex input voltage and current, and V_{out} and I_{out} as complex output voltage and current, the behavior in the frequency domain of a single stage of a polyphase network can be expressed in a similar way as that of a two-port [Ging EC73, Galal TCASII00], as introduced in Section 2.4.2.3:

$$\begin{bmatrix} V_{in} \\ I_{in} \end{bmatrix} = \mathbf{T} \begin{bmatrix} V_{out} \\ I_{out} \end{bmatrix} \tag{4.15}$$

The transmission matrix representation $\mathbf{T}$ for one stage of the complex filter is given by:

$$\mathbf{T} = \begin{bmatrix} t_{11} & t_{12} \\ t_{21} & t_{22} \end{bmatrix} = \frac{1}{1 + jsRC} \begin{bmatrix} 1 + sRC & R \\ 2sC & 1 + sRC \end{bmatrix} \tag{4.16}$$

The transmission matrix for a multi-stage network is obtained by matrix multiplications of the chain matrices of the consecutive stages, as already explained in (2.29) of Section 2.4.2.3.

For a three stage polyphase network, the complex transfer function calculated analytically from the transmission matrix for a zero output current is given by

---

[1]In a standard two-port description, a negative sign is used for I_{out} [Chua 87]. The convention used in this work is that the current I_{in} flows into the circuit and I_{out} flows out. For I_{out} this means the current flows in an inverse direction when compared to the 'normal' two-port description.

[Galal TCASII00]:

$$\frac{V_{out}(s)}{V_{in}(s)} = \frac{(1 + jsR_1C_1)(1 + jsR_2C_2)(1 + jsR_3C_3)}{1 + as + bs^2 + cs^3} \tag{4.17}$$

where

$$a = R_1C_1 + R_2C_2 + R_3C_3 + 2(R_1C_2 + R_2C_3 + R_1C_3)$$
$$b = R_2R_3C_2C_3 + R_1R_2C_1C_2 + R_1R_3C_1C_3$$
$$+ 2(R_1R_3C_2C_3 + R_1R_2C_2C_3 + R_1R_2C_1C_3)$$
$$c = R_1R_2R_3C_1C_2C_3$$

This formula can be used to calculate the frequency response of the polyphase network. However, to obtain the needed values for R's and C's to realize some desired frequency response, an iterative procedure has to be used. The advantage of using this analytical approach as compared to a circuit simulation-based approach thus is very doubtful, since a circuit-evaluation of a polyphase network is not substantially slower than an evaluation of the calculated transfer function. Other elements like the inclusion of loading effects, mismatch effects and total design time also lead to a preference of circuit simulation to the use of analytical formulae. This will be pointed out in the following paragraphs.

Loading effects & input impedance The effect of an output load on the network behavior can not be readily seen using the transmission representation as described above. To use the polyphase network as described by (4.17), no current is to be drawn out of the polyphase network by the circuitry following it. Therefore, the input impedance of this circuitry has to be high as compared to the impedance of the network itself. A typical example of an output circuit is a voltage buffer, that is used to compensate for the inherent losses of the passive polyphase network, e.g. a MOST single stage amplifier. Then, the load of the polyphase network can be considered capacitive. To look at the influence of this load on the performance of the polyphase network, the analytical approach as described above is rather useless. However, starting from the transmission representation of a two-port (4.16) and using some calculations, a transfer characteristic incorporating the loading effect can be obtained. Also using the elements of **T** the input impedance of a loaded polyphase network can be calculated. This input impedance is useful for the design of the input buffer of the polyphase network as already mentioned in Section 2.4.2.5.

The following method is used to include the loading effects [Chua 87] in the calculation of the complex transfer function:

1 From the **T**-matrix the input impedance of a polyphase network loaded with an impedance Z_2 can be derived:

$$Z_{11} = \frac{t_{11}Z_2 - t_{12}}{t_{21}Z_2 - t_{22}} \qquad (4.18)$$

2 The **T** or *transmission* matrix is converted to the **Z** or *impedance* matrix, using the following conversion scheme:

$$\mathbf{Z} = \begin{bmatrix} z_{11} & z_{12} \\ z_{21} & z_{22} \end{bmatrix} = \begin{bmatrix} \frac{t_{11}}{t_{21}} & \frac{-det(\mathbf{T})}{t_{21}} \\ \frac{1}{t_{21}} & \frac{-t_{22}}{t_{21}} \end{bmatrix} \qquad (4.19)$$

with the matrix elements t_i as defined in (4.16). The element z_{11} of the **Z**-matrix is the input impedance of the unloaded polyphase network.

3 The voltage transfer function of a polyphase network driven by a source V_{in} with a series impedance Z_1, and loaded with and impedance Z_2 (as shown in Fig. 4.5) is calculated using the elements from the **Z**-matrix as follows:

$$\frac{V_{out}(s)}{V_{in}(s)} = \frac{z_{21}Z_2}{(z_{11} + Z_1)(z_{22} + Z_2) - z_{12}z_{21}} \qquad (4.20)$$

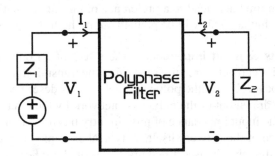

Figure 4.5.   2-port representation of a loaded polyphase network

The results of the calculations for a three stage polyphase network, are too cumbersome to include here. Moreover, they do not give much insight in the effects that occur. A better way to gain some insight is to plot the calculated formulae. Therefore, some plots are given that visualize the influence of the loading effect and show the influence of the series impedance of the source.

Fig. 4.6 shows the complex voltage transfer characteristic and input impedance of a loaded polyphase network, with an ideal voltage source as input ($Z_1 = 0$ Ohm in Fig. 4.5). The effect of a non-zero input source series impedance, is

Table 4.1. R&C Values of the three stage polyphase network used for simulations

resistor	value [Ω]	capacitor	value [fF]
R_1	295	C_1	800
R_2	200	C_2	800
R_3	110	C_3	800

demonstrated in Fig. 4.7 for a capacitive load impedance Cload= 0.5pF. The values of capacitors and resistors of the three stage polyphase network used for these calculations are given in Table 4.1.

Another topic, that can be considered a 'loading effect' is the loading by consecutive stages of the polyphase network. To demonstrate this effect, a very simple comparison is done. Calculations performed on the polyphase network as depicted in Table 4.1 are compared to calculations for a polyphase network with resistors in reverse order, i.e. $R_1 = 110\ \Omega, R_2 = 200\ \Omega, R_3 = 295\ \Omega$. The plots in Fig. 4.8 show the results. The reversed resistance order results in a minor suppression of negative frequencies, and an input impedance that drops to a lower value at a lower frequency. This lower input impedance calls for an input buffer of the polyphase network with an even lower output impedance, causing a higher power consumption.

From the discussion above it is clear that the circuitry driving and following a polyphase network has an influence on the transfer characteristic that is not negligible at all. The influence of the order of the consecutive stages on the frequency response and on the input impedance has also been demonstrated. In practice, the output impedance of any buffer used as polyphase network driver has a complicated behavior in the frequency domain. Since the source impedance of the driver will also influence the transfer characteristic and since the real behavior of this impedance is difficult to include in the analytical approach, the analytically modeled transfer characteristic is not always reliable.

In fact, the final design can only be characterized accurately using a circuit simulator. Due to the entanglement described above, minor changes in the polyphase network also lead to an iteration in the design of the building blocks interacting with the polyphase network. A direct consequence is that the (small) simulation time benefit of the analytical approach is lost completely. Therefore, the proposed work flow does not start from the analytical approach but uses circuit simulations from the beginning of the design.

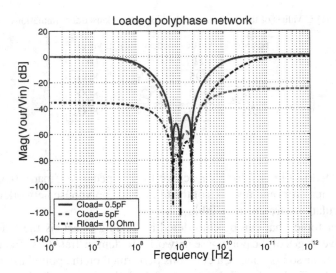

(a) Voltage transfer characteristic

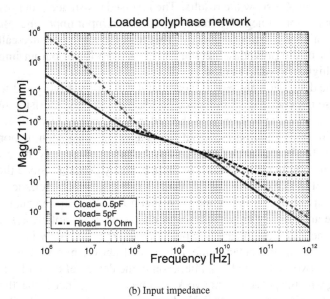

(b) Input impedance

Figure 4.6.   Effect of output loading on PF, $Z_1 = 0$

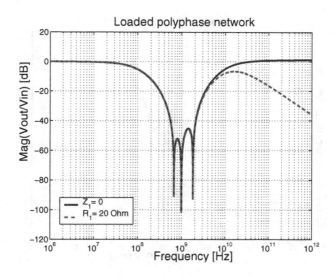

(a) Voltage transfer characteristic

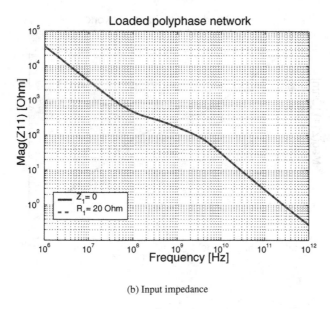

(b) Input impedance

Figure 4.7. Effect of input source resistance on PF, Cload = 0.5pF

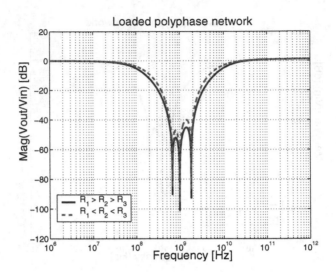

(a) Voltage transfer characteristic

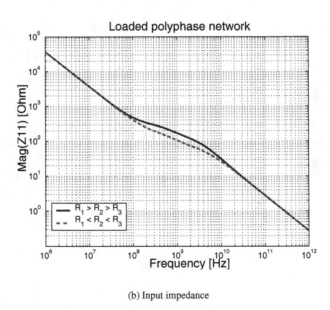

(b) Input impedance

Figure 4.8.   Effect of resistor order on PF, Cload = 0.5pF, $Z_1 = 0$

Circuit simulation As stated in the previous paragraph, final simulation is done preferably using a circuit simulator. Since this kind of simulators can only work with real signals, a specific simulation approach has to be used to simulate a complex filter. For a full evaluation in an AC simulation of a complex network, four instances ($PF_{1...4}$) of this network are needed. The theoretical explanation, followed by the practical implementation of the test bench is given in this paragraph.

The complex equivalent of a simple single-frequency sine wave is a complex signal defined as:

$$X_{PF1} = A \cdot e^{j\omega t} \tag{4.21}$$

and the complex output can be calculated as a complex sum of two real signals, as defined by (4.14). In an AC simulation, this complex sum can also be obtained as a direct sum of two real signals, by exploiting the fact that a polyphase network is in essence a linear system and thus (4.9) can be used. Thus, the complete second term, including the j-factor, of (4.14) can be obtained as the quadrature output y_Q in Fig. 4.4 with the following complex signal as input:

$$X_{PF2} = j \cdot A \cdot e^{j\omega t} \tag{4.22}$$

The complex output signal is then obtained as:

$$Y_{pos} = y_I^{PF1} + y_Q^{PF2} \tag{4.23}$$

The network can now be evaluated for positive frequencies but not yet for negative, since in the circuit simulator only positive values for ω are allowed. This already necessitates the instantiation of two copies of the polyphase network to be simulated, with different complex input signals.

For the evaluation of the network for negative frequencies two more instances of the polyphase network are needed. The input signals applied to these two networks are: $X_{PF3} = Ae^{-j\omega t}$ and $X_{PF4} = j \cdot Ae^{-j\omega t}$. The output is then obtained as:

$$Y_{neg} = y_I^{PF3} + y_Q^{PF4} \tag{4.24}$$

Hereby, the question of how to obtain a complex output vector using an AC simulation with real output signals only, is solved. But the question remains how to apply the input signals as defined above. This problem is solved quite easily by writing down the complex signals as a sum of cosines:

$$Ae^{j\omega t} = A(\cos \omega t + j \cdot \cos(\omega t - 90°)) \tag{4.25}$$

$$j \cdot Ae^{j\omega t} = A(\cos(\omega t + 90°) + j \cdot \cos \omega t) \tag{4.26}$$

$$Ae^{-j\omega t} = A(\cos \omega t + j \cdot \cos(\omega t + 90°)) \tag{4.27}$$

$$j \cdot Ae^{-j\omega t} = A(\cos(\omega t - 90°) + j \cdot \cos \omega t) \tag{4.28}$$

Comparing these equations with the definition of the complex input vector X, as given in (4.13), it becomes clear that we only have to apply a cosine input source with the appropriate phase to each input of all four instances of the polyphase network.

In Fig. 4.9 the complete test bench for an AC evaluation of a polyphase network used as a complex filter is given. The top part shows the part of the circuit that is used to simulate the AC network behavior for positive frequencies whereas the bottom part shows the part of the circuit that simulates the AC behavior for negative frequencies.

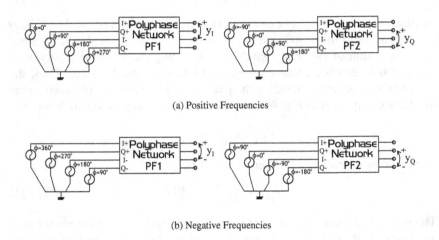

(a) Positive Frequencies

(b) Negative Frequencies

Figure 4.9.   AC simulation of a Polyphase Network used as complex filter

Using this test bench, the behavior of a polyphase network can be evaluated in the complex frequency domain using only one single AC simulation. Thus, no postprocessing steps using e.g. Matlab [Matlab], are necessary.

Mismatch effects   The inclusion of mismatch in a polyphase network is discussed thoroughly in Section 2.4.2 for its effect on the quadrature accuracy. As explained there, the effect of mismatch on the complex filter characteristic can be estimated using formula (2.34), expressed as a maximum attainable suppression level for a certain mismatch percentage on resistors and capacitors. For a more accurate study of the influence of mismatch, the simulation approach as shown in Fig. 4.9 can be used, utilizing the parametrized base-cell of the polyphase stage including mismatch effects, as defined in Fig. 2.29.

Influence of harmonics In Section 2.4.2.8 it is indicated that the presence of higher harmonics in the differential oscillator signal driving a polyphase network to obtain a quadrature signal can deteriorate the image rejection of an upconversion system in which this quadrature signal is used. Here, higher harmonics are filtered out by the polyphase network. Therefore, the image rejection will be less influenced by the harmonics present in the input signal.

Influence of quadrature inaccuracy Another non-ideality that can deteriorate the filter characteristics of the polyphase network, is the presence of quadrature inaccuracy in the input signal of the network. As calculated in Section 4.1.2.2, a quadrature square wave has an asymmetric frequency spectrum with harmonics alternating between positive and negative frequency. As is mostly the case, quadrature inaccuracy leads to some kind of "leakage" in the frequency domain. In this particular case, there will be some positive frequency content for harmonics normally present only at a negative frequency, and vice versa.

The result is that any $(4n+1)th$ harmonic will be subject to a power loss compared to the ideal case, since their negative frequency component present due to the quadrature mismatch will be attenuated by the polyphase network. Adversely, any $(4n-1)th$ harmonic will be filtered out less as compared to the ideal case, since their positive frequency component present due to the quadrature mismatch can pass unattenuated through the polyphase network.

Analytical calculations vs. Circuit Simulation Nowadays, circuit simulation is still regarded as the most reliable method for final circuit validation. Therefore, the polyphase network will eventually be simulated and possibly fine-tuned using circuit simulations. For the designs presented in this work, the polyphase filter has never been designed using the analytical method. As is demonstrated above, it is a good way to gain insight and possibly also to get a good starting point for the resistor and capacitor values. However, the time needed for an AC simulation of a polyphase network is almost comparable to the time needed for analytical calculations. Added to that, there will always be some difference in the results of the analytical approach and the simulation which are mostly due to the input/output buffering. Thus, almost no time is gained by using the analytical approach.

4.1.2.4 Complex Filtering in Reality

The combination of the complex square wave property explained in the previous paragraph and the asymmetric suppression of negative frequencies in the polyphase network explained above, leads to an active suppression of the $(4n-1)th$ harmonics of a quadrature oscillator signal. In this way, the intrinsically non-linear oscillator signal is linearized by a complex filtering through the polyphase network. To verify this theory, different test chips have been designed, all incorporating a polyphase network. In Fig. 4.10 a comparison is shown between the simulated and measured complex transfer characteristic of one of the polyphase networks implemented on chip. The values for capacitors and resistors of this particular polyphase network are given in Table 4.2.

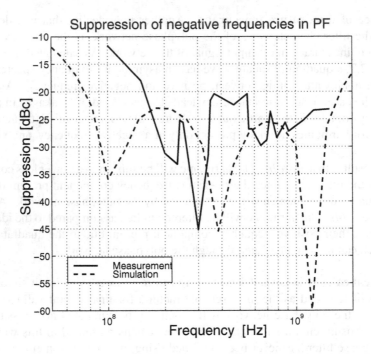

Figure 4.10. Simulated and measured transfer characteristics of a PF with component values given in Table 4.2

Table 4.2. R&C Values of implemented three stage polyphase network

resistor	value [Ω]	capacitor	value [pF]
R_1	150	C_1	0.88
R_2	150	C_2	2.67
R_3	150	C_3	9.96

This polyphase network has been integrated as part of a CMOS two channel upconvertor to actively linearize the on-chip ringoscillator [Borr JSSC99]. The linearized ringoscillator is presented in full detail in Section 4.1.4.

4.1.3 Method of Waveform Construction

Complementary to the use of a polyphase network to obtain a local oscillator signal with lower harmonic content, a waveform-construction method can be used. Instead of generating a square wave, a wave is constructed that has no fifth harmonic by construction [DeRa ESSCIRC99]. Fig. 4.11 shows the construction procedure. In Fig. 4.11(a), a normal square wave is shown, with its

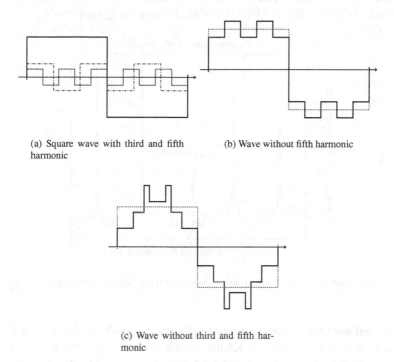

(a) Square wave with third and fifth harmonic

(b) Wave without fifth harmonic

(c) Wave without third and fifth harmonic

Figure 4.11. Waveform construction

square components at the third and fifth harmonic frequency. The amplitude of the third and fifth harmonic of a square wave is $1/3$ resp. $1/5$ of the amplitude of the first harmonic. However, if we would want to subtract only the third

harmonic in the Fourier spectrum of the square wave, we would need to create
a sine wave with an amplitude of $4/(3\pi)$ given a square wave with a 1 V
amplitude. If a square wave is subtracted, the amplitude of the square wave
being subtracted simply has to be $1/3$ or $1/5$ of the amplitude of the original
square wave. This method is used in Fig. 4.11(b) to construct a square wave
without fifth harmonic, and in Fig. 4.11(c) to construct a square wave without
third nor fifth harmonic. Using this technique also the frequency components
at the harmonics of the subtracted square wave are suppressed. E.g. the 15th
harmonic of the constructed wave of Fig. 4.11(b), being the third harmonic of the
subtracted waveform, is also suppressed as compared to the normal square wave.
In Fig. 4.12 the Fourier spectrum of the constructed waveform of Fig. 4.11(b)
is shown. It is clear that the fifth harmonic is completely suppressed.

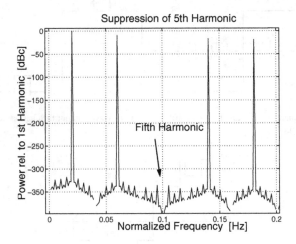

Figure 4.12.   Theoretical suppression of fifth harmonic by construction

A real example   For the broadband system introduced in Section 4.1.1,
a clean oscillator signal is needed, with low harmonic content up to a frequency of 1 GHz. A digital circuit implementation could be used to generate
a wave form without fifth harmonic. For this application, the highest oscillation frequency for which no fifth harmonic content can be tolerated would
be 200 MHz, corresponding to a fifth harmonic at 1 GHz. Since the digital
logic would have to generate a signal with a time resolution of 1 GHz it
would be working at a clock frequency of (at least) 1 GHz. To generate the
waveform of Fig. 4.11(c), an internal clock frequency of 3 GHz would be
needed. This frequency is maybe feasible in the newest (expensive) CMOS
technologies, but the use of a polyphase network to filter out the third-order
harmonic still seems a more realistic approach.

To investigate in practice the use of a polyphase network combined with wave form construction, measurements have been performed on a test chip. The test chip itself is presented in the next section, but the measurement results for the wave form construction are given here. The input is a quadrature signal with a wave form as shown in Fig. 4.11(b) constructed using a digital data generator. Due to restrictions on the maximum clock frequency of the data generator, the highest quadrature square wave that can be generated has a base frequency of 50 MHz. This quadrature signal is applied to an on-chip polyphase network, that is used to filter out the third harmonic of the quadrature square wave. Fig. 4.13 shows the frequency contents of the output signal of the data generator and of the output signal of the polyphase network, for a 50 MHz LO oscillator signal. For sake of completeness, the power in the fifth-order harmonic of a "normal" square wave coming out of the data generator is also shown in the figure. The graph clearly demonstrates that the fifth-order harmonic is suppressed by construction and that the third-order harmonic is suppressed by the polyphase network. Moreover, it can be seen that also the seventh harmonic is suppressed by the polyphase network. This can be explained by looking at the complex frequency spectrum of the square wave in Fig. 4.3 and by noting that the seventh harmonic of the 50 MHz LO signal is situated at 350 MHz, which is still in the suppression band of the polyphase network.

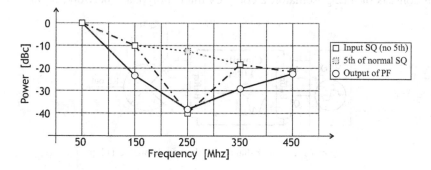

Figure 4.13. Measurements to validate wave form construction method

As final remark, it should be noted that no digital implementation of a wave form constructor has been made in the context of this work. Therefore, no estimate of the power consumption or area usage of such an implementation can be given at this point. However, recent publications show that the implementation of multi clock-phase based frequency synthesizers is feasible [Spat JSSC02] and this method could be used to generate the wave forms as studied in this section.

Table 4.3.   R&C Values of the three stage polyphase network of Fig. 4.15

resistor	value [Ω]	capacitor	value [pF]
R_1	100	C_1	1.32
R_2	100	C_2	4.0
R_3	100	C_3	14.9

4.1.4   Chip Implementations of LO Linearization

To test the theory described in the previous sections, some chip implementations have been made. A first test chip encompasses a complex filter, an on-chip ringoscillator and a quadrature mixer block to convert the quadrature signal into a single-ended one, ready for measurements on a standard spectrum analyzer. Thus, a single-channel transmitter is integrated on chip. In a second test chip, two such transmitters are combined to investigate the integratability of multi-channel transmitters for cable broadcast applications.

4.1.4.1   LO Linearization using a Complex Filter

The block diagram of a first test chip is shown in Fig. 4.14 [DeRa ECCTD99]. It consists of a ringoscillator, a complex filter (polyphase network, input and

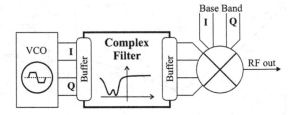

Figure 4.14.   Block diagram of the one channel test chip

output buffers) and a quadrature upconvertor block. Fig. 4.15 depicts the detailed circuit schematic of the complex filter implemented in the test chip. The input buffers are four identical MOST source followers *(block I)*. The polyphase network itself *(block II)* has three stages, with values for capacitors and resistors as given in Table 4.3.

The design of the polyphase network is based on the zero-voltage output approximation, thus necessitating an output buffer with a low input impedance. As shown in Fig. 4.15, this is realized in *block III* by connecting the output of the polyphase network to the source of nMOS transistor M_c that is in a common-gate configuration. The voltage at the input of the polyphase network

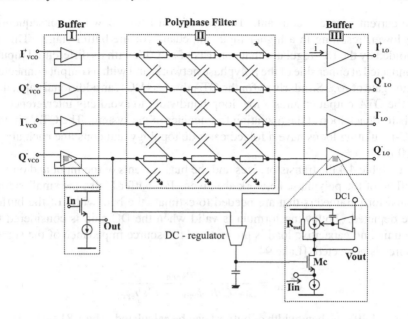

Figure 4.15. Circuit schematic of the complex filter

is converted into a current flowing out of the polyphase network. The parallel impedance of the current source biasing M_c must be as high as possible, since any AC-current flowing into this impedance is lost. The AC-current is converted to a voltage output at the drain of transistor M_c. The gain of *block III* is limited by resistor Rout of Fig. 4.15, whereas the DC-output level is set to the value of DC1 by a local feedback loop. The gain and mismatch of the OTA used in the feedback loop will determine how accurately the DC level at the output matches DC1. Since the exact voltage at the output is not crucial for the performance, the specifications of the OTA are quite relaxed. The only real important specification is the position of the first pole; since all frequencies below the GBW of the OTA are suppressed, it has to be reassured that the GBW is lower than the minimal frequency of interest (here around 100 MHz). In this design, the GBW was made lower than 50 MHz by addition of a capacitance at the output of the OTA.

Another local feedback loop tries to minimize the DC voltage drop over the polyphase network, to avoid any influence of the DC-coupling of the polyphase network on the input impedance of the output buffer (III). By minimizing the DC voltage drop, the DC current flowing through the network is also minimized. The input impedance of the output buffer of the polyphase network is in first order equal to $1/gm_{Mc}$. Any DC current flowing through the network, would result in a lower DC current flowing through M_c since the bias current through the current source is constant. The gm of transistor M_c would consequently be lower, resulting in a higher input impedance of the buffer stage. This is avoided by the DC-regulator that consists of an OTA with its differential inputs connected at either side of the polyphase network, and with its output connected at the gate of M_c. Similarly as for the OTA that sets DC1, an additional capacitor at the OTA output ensures a low loop bandwidth to avoid any interference of this feedback loop at the operating frequencies of the system. The OTA's in the DC-regulating loops have a folded cascode topology and consume each about 200 μA.

In Table 4.4 the transistor sizes and the bias currents of the input and output buffers of the polyphase network are given. For buffer I, also the small-signal transistor parameters that are needed to estimate the bandwidth of the buffer are depicted. Following formula is valid when the DC gain is considered to be unity, no capacitive load is present and the source impedance of the signal source is neglected ([Lake 94]):

$$f_{3dB} = \frac{gm_{src}}{2\pi(Cdtot_{curr} + Csb_{src})} \tag{4.29}$$

Using (4.29), the bandwidth of buffer I can be calculated to be 8.81 GHz. However, in reality the buffer drives a polyphase network. As a first simplification this load is estimated to be the DC resistance of the polyphase network, being 450 Ω. If also the drain-source impedance of the current source is included, the DC gain turns out to be 0.9. The gate-source capacitance Cgs_{src} therefore has to be taken into account for 10% of its value. This results in a bandwidth of 7.8 GHz. Obviously, this is only a rude estimate since the load impedance is in reality much more complex than a simple resistance (see Fig. 4.6). However, it should be clear that the bandwidth of the buffer is not a limiting factor here. In this design, the key performance parameter of buffer I is its DC output impedance, equal to $1/gm_{src}$.

The ringoscillator is built up out of four differential invertors, as shown in Fig. 4.16. This topology provides us with a differential quadrature signal for free. The conditions for a stable oscillation are defined as the Barkhausen Criterion, being an open loop phase shift of 180° and an open loop gain > 1. Theoretically this can be obtained in a ringoscillator with two stages, but then care must be taken to avoid a non-oscillating meta-stable condition of the oscillator. Therefore, mostly a three stage ringoscillator is used when high-frequency/low-power operation is wanted. This is readily explained as follows:

- The oscillation frequency is inverse proportional to the invertor delay time and the number of stages;

- The delay time is inverse proportional to the bias current, thus power usage per stage.

Table 4.4. Transistor sizes and bias currents of Fig. 4.15

Buffer I	I_{bias}	2.5 mA
	$(W/L)_{src}{}^a$	$(395/0.5)\mu m$
	$(Vgs - Vt)_{src}$	0.17 V
	gm_{src}	21.7 mS
	Csb_{src}	157 fF
	Cgs_{src}	500 fF
	GBW (calculated)	7.8 GHz
	$(W/L)_{curr}{}^b$	$(260/1)\mu m$
	$(Vgs - Vt)_{curr}$	0.38 V
	$(gm)_{curr}$	11.4 mS
	$(Cdtot)_{curr}$	235 fF
	$(gds)_{curr}$	10.6 mS
Buffer III	$(W/L)_{Mc}$	$(190/0.5)\mu m$
	$(W/L)_{currtop}{}^c$	$(65/0.5)\mu m$
	$(W/L)_{currbtm}{}^d$	$(160/1)\mu m$
	Resistor R_{out}	500 Ω
	I_{bias}	1.2 mA

[a] Source follower; nMOST device
[b] Current source; nMOST device
[c] Regulated current source on top; pMOST device
[d] Current source bottom; nMOST device

This means that the oscillation frequency is inverse proportional to the power usage and to the number of stages. To obtain a quadrature signal directly as output from the ringoscillator, the 180° open loop phase shift of the circuit must be divided by two or a multiple of two. Since a two stage ringoscillator is somewhat unreliable, a four stage ringoscillator is the obvious choice. The oscillation frequency of the oscillator can be tuned by changing the delay time of the invertors. How this is done, is explained based on the schematic of the differential invertor used.

The transistor schematic of a single invertor of the ringoscillator is given in Fig. 4.17. The core of the invertor is formed by the differential pair M1. The bias current of M1 is supplied by current source transistor M4. The load of the differential pair consists of the pMOST diode pair M2 and the transistors M3. The delay time is proportional to $Rout/gm_{M1}$, with $Rout$ the (resistive part of the) impedance at the output node. For normal operation, the voltage $Vbias2$ is set equal to the power supply in order to switch off transistors M3. Then, $Rout$ is in first order equal to $1/gm_{M2}$. Since for a MOS-transistor in

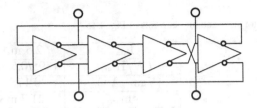

Figure 4.16.   Four stage differential ringoscillator

saturation $gm \propto \sqrt{I}$, the delay time is proportional to $1/I$. The voltage gain, which is determined by gm_{M1}/gm_{M2}, remains approximately constant. In this way a broad frequency range can be obtained by varying $Ibias$. In practice, the gate voltage $Vbias1$ is used to change the bias current and tune the oscillator frequency, thus making it a *voltage-controlled* oscillator.

For low frequency operation, a low tuning voltage for $Vbias1$ and consequently low currents for $Ibias$ are selected. The output resistance $Rout$ will then be high. To make it even higher, the load transistors M3 can be switched on by adjusting $Vbias2$ accordingly. Then, transistors M3 take over the current of the diodes M2, raising the impedance at the output nodes, thus raising the delay time of the invertor. This way, lower oscillation frequencies can be obtained than with the diode loads alone.

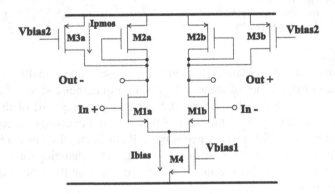

Figure 4.17.   Differential invertor for the ringoscillator

Fig. 4.18 shows the measured tuning range as a function of both biasing nodes. A very wide tuning range from 55 MHz to 1200 MHz, being more that a decade has been achieved. The excess tuning range is a safety margin for technological variations. The lowest frequencies have been measured with M3 in the on-state.

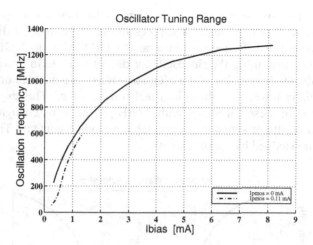

Figure 4.18. Measurend tuning range of the ringoscillator

Table 4.5. Transistor sizes of differential invertor of Fig. 4.17

$(W/L)_{M4}$	$(180/1)\mu$m
$(W/L)_{M1}$	$(90/0.5)\mu$m
$(W/L)_{M2}$	$(97/0.5)\mu$m
$(W/L)_{M3}$	$(90/4)\mu$m

The phasenoise of the ringoscillator measured at an offset of 600 kHz from the carrier is -94 dBc/Hz at an oscillation frequency of 1.2 GHz. Over the whole frequency range, the phasenoise varies between -87 dBc/Hz and -94 dBc/Hz. Compared to the LC-tank type oscillator presented in Section 3.4.6 or the one presented in the next section, this phasenoise is very high. Together with the high power consumption, this is the major drawback of ringoscillators.

In Table 4.5 the sizes of the transistors of the differential invertor used in the ringoscillator are given. They have been obtained by an equation-based optimization towards minimal power consumption for a specified frequency range of operation, with first-order models for the transistors (Spice Level 2). The length of transistor M3 has been made large to obtain a high output resistance and thus realize a low oscillation frequency when this transistor is switched on. The bias current I_{bias} in Fig. 4.17 varies between 0.2 mA and about 5 mA to cover the whole frequency range, as can be seen in Fig. 4.18. The resulting power consumption from a 3.3 V power supply thus varies between 2.64 mW and 66 mW for the complete ringoscillator.

The transfer function of the polyphase network regarded as a complex filter is simulated using the method described above in Section 4.1.2.3. In Fig. 4.19

the simulated relative suppression of the negative frequencies compared to the positive frequencies is plotted. A suppression of more than 20 dB is to be expected for negative frequencies in a range of 100 MHz to 1 GHz. Since the invertors constituting the ringoscillator do not provide full swing square wave signals, the harmonics of the oscillator signal as generated on chip will be substantially lower than those in a real square wave. This explains why a suppression of these harmonics with 20 dB proved to be enough to obtain a linearized oscillator signal with acceptable harmonic content. This will be further acknowledged by measurements.

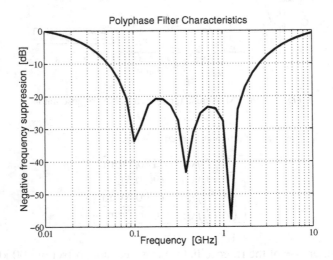

Figure 4.19.   Simulated complex transfer characteristic of the polyphase network

First, a simulated result is shown. Fig. 4.20 demonstrates the suppression of the third harmonic frequency by the complex filter. In Fig. 4.20(a) the simulated frequency spectrum of the input signal of the polyphase network is shown, and in (b) the simulated frequency spectrum at the output of the polyphase network is given. The third-order harmonic is clearly reduced by the polyphase network, acknowledging in simulation the theory derived above.

The test-chip is realized in a $0.5\mu m$ standard CMOS technology. In Fig. 4.21, a chip photograph is given. From left to right, the four stage ringoscillator *(1)*, the polyphase network with its input and output buffers *(2)* and the quadrature mixer block *(3)* can be distinguished.

The measurements of the output signal of this test chip are summarized in the plot of Fig. 4.22. For a given oscillation frequency of the on-chip ringoscillator, the power of the higher harmonics relative to the power measured in the first harmonic are plotted. Above an oscillation frequency of 100 MHz, all harmonic

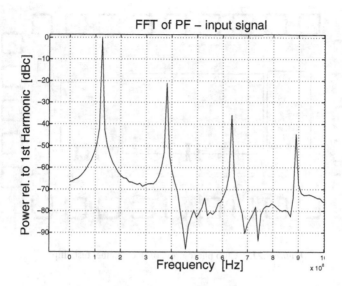

(a) Before polyphase network

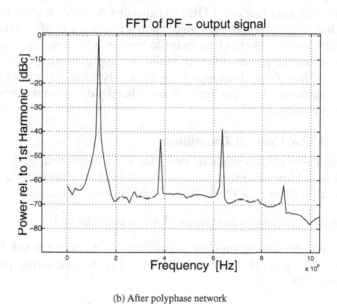

(b) After polyphase network

Figure 4.20. Simulated third-order harmonic suppression of ringoscillator signal

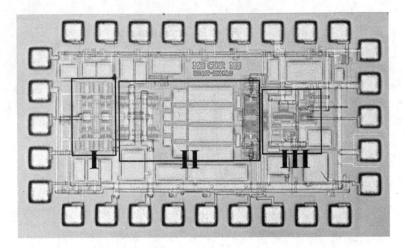

Figure 4.21.   Chip photograph of one channel test chip

components are less than -39 dBc. It can also be seen that the power of the third-order harmonic is lower than that of the fifth-order harmonic, up to a frequency for the fifth-order harmonic of 1 GHz and also that the seventh-order harmonic is lower than the ninth-order harmonic, up to a frequency of 900 MHz for the ninth-order harmonic. As can be seen in the Fourier expansion of a square wave (4.10), normally the power of the harmonics is simply inverse proportional to their order. Thus, the active suppression by a polyphase network of the *(4n-1)th*-order harmonics of a quadrature wave is hereby demonstrated once again with measurements.

4.1.4.2   A Two Channel Transmitter

A second test chip consists of two identical transmitters as depicted in Fig. 4.14 of which the output current is combined on chip using a cascode transistor stage [Borr ISSCC99]. This test chip has two versions:

- A first one with an on-chip ringoscillator in each channel;

- A second one without on-chip ringoscillators, but with the possibility to connect the inputs of the on-chip complex filter to an external quadrature LO generator.

The second version of the test chip is used to accurately test the suppression of third-order harmonics by the polyphase network. The plot of Fig. 4.10 is based on measurements on this version of the test chip. Also the results of Fig. 4.13 are obtained using this test chip. The values of the capacitors and resistors of the polyphase network of the two channel transmitter are given in Table 4.2.

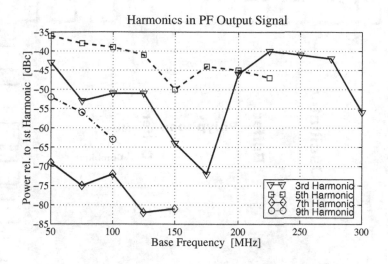

Figure 4.22. Measured harmonics in the output signal

Compared to the component values of the single-channel transmitter given in Table 4.3, higher resistance values and lower capacitance values are used for the two channel transmitter. This is due to a downscaling of the output power per channel and a resulting reduction of the input capacitance of the mixer block. An optimization over all building blocks of the complex filter of Fig. 4.15 then resulted in a power reduction in the buffers and an area reduction in the polyphase network by a reduction of the total capacitance value. Fig. 4.23 shows a chip photograph of the two channel transmitter with on-chip ringoscillators. Clearly, two identical parts can be discerned on the chip corresponding to the two single-channel transmitters.

In Table 4.6 the measured specifications of the two channel transmitter are summarized.

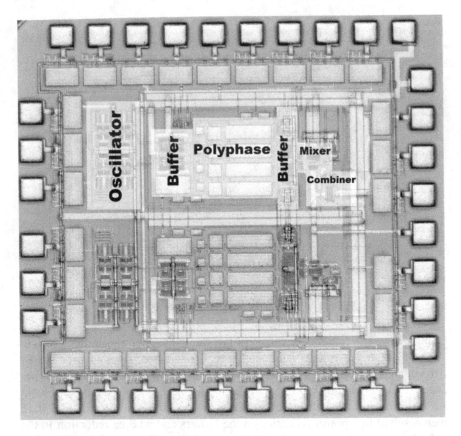

Figure 4.23. Chip photograph of the two channel transmitter

4.2   Oscillators for RF Frequencies

The design of RF oscillators and synthesizers in a standard CMOS technol-ogy, has only made it into commercial products in recent years ([Silabs]). In research, the pursuit towards ever higher operation frequencies for the com-plete synthesizer ([Hung JSSC02]) and particularly for the oscillator ([Wang ISSCC99, Klev ISSCC99, Wang ISSCC01, Tieb ISSCC02]) still proves to be a source of many publications. Of course, research is without value without possible applications. For the research domain of high-frequency, low-phase noise synthesizers and oscillators, the applications are in the field of:

■ High-speed optical communications; This type of application needs devices with low-phase noise oscillators to minimize the total number of active devices (*e.g. repeaters*) within the passive optical network (PON). The lower the phase noise, the longer an optical fiber can be before loss of data occurs;

Table 4.6. Specifications of two channel transmitter

Technology	$0.5\mu m$ CMOS
Frequency range	100 MHz- 1100 MHz
Number of parallel channels	2
Topology	Direct Conversion
Power Supply	3.3 V
Output Power (per channel)	-16 dBm in 75 Ω
Harmonic Distortion	$<$-40 dBc
Interchannel Intermodulation	$<$-47 dBc
Current Drain (per channel)	28.2 - 43.8 mA
Total Chip Area	1.25 x 1.25mm^2
Noise Floor	-139.8 dBc/Hz

- High-bandwidth (fixed point) wireless communications in (un)licensed bands; Here low phase noise is important to maximize the bandwidth efficiency of the telecommunication link.

In this section the design of a high-frequency voltage-controlled oscillator (VCO) is presented. A first subsection discusses some general oscillator design issues. In a second subsection, the topology is presented and an overview is given of the different building blocks of the VCO including a discussion of the influence of their parasitics on the oscillator's performance. A third subsection explains in which case the proportionality between power and phase noise that is normally valid for LC-type VCO's, no longer holds. In a fourth subsection the measurement results of a 17 GHz VCO that has been designed using the in-house VCO design tool CYCLONE and manufactured in a $0.25\mu m$ CMOS technology are presented and the final subsection compares this VCO with recent work.

4.2.1 Oscillator Design Issues

In the design of fully integrated oscillators in general and in the design of oscillators operating at very high frequencies in particular, some specific issues have to be taken into account:

- The skin effect is related to the frequency. This means that, in the design of the on-chip coil, an increase of the parasitic series resistance becomes more and more important. The result is an enlargement of the phase noise of the oscillator;

- The phase shift of the gain-cell at the working frequency of the oscillator, on top of the gate resistance itself, becomes important. Any additional phase shift introduced by the physical gate resistance has to be taken into account in the design. This is done here by introducing a gate resistance in series with the gate-node of the MOST for simulation purposes;

- The resistance of both gain cell and diode cell is becoming more important than the series resistance of the coil. The reasons for this are:

  - The low value needed for the inductance, resulting in a small coil with small length and *relatively* low series resistance;

  - The low value needed for the capacitance, leading to a low total diode length and a high resistance, since the diode resistance is inverse proportional to the total diode length;

  - The small number of needed gain cell blocks, resulting in a high total gate resistance (see (4.32)). A small number of blocks is needed since for a given inductance L, a high oscillation frequency results in a low needed G_m (see (4.31)).

- Due to the previous issue, a full optimization of the VCO circuit minimizing the total effective resistance is probably the most reliable method to obtain an optimally designed circuit;

- Special attention has to be given to a high layout symmetry in both floorplan and building blocks, to avoid the introduction of any additional asymmetry in the oscillation waveform. This would cause unwanted extra noise upconversion, as depicted in [Haji JSSC98].

4.2.2   Building Blocks

The topology for the VCO that has been designed is shown in Fig. 4.24. It has a very simple, but very symmetrical topology to avoid unexpected problems due to high-frequency effects not taken into account by the circuit simulator. The design equations that are needed for the sizing of this topology are given in Section 3.4 and some are repeated here when needed. In the schematic following important building blocks of the VCO can be discerned:

- An inductor: here only symmetrical octagonal coils are used;

- A varactor diode: it has a custom layout;

- A gain cell: this cell also has a custom layout;

- A bias transistor pair.

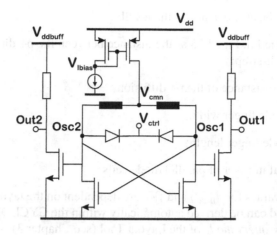

Figure 4.24.   Topology of the integrated RF-VCO

The symmetrical octagonal coil with central tap has some advantages that are already mentioned in Section 2.3.3.3:

- A self-oscillation frequency twice as high as that of the same coil used in a single-ended topology[Nikn ESSCIRC99];

- A lower phase noise than a square coil with the same inductance value [Most TCASII01, Cran PhD97].

The layout of the coil is generated using an automated layout generator and its complex impedance is simulated using FastHenry. Both of these steps are performed within the automated VCO design tool CYCLONE presented in Section 3.4. The series resistance of the coil has a direct influence on the phase noise performance of the VCO (see (3.9)), and therefore it should be kept low. In this design, the on-chip inductor uses all 4 available metal layers connected in parallel to reduce the sheet resistance of the coil.

The custom layout of the varactor diode cell is shown in Fig. 4.25. The varactor diodes are implemented using several blocks connected in parallel. As indicated in the drawing, one diode block consists of two diode fingers *(p+ diffusion)* surrounded by nwell-straps *(n+ diffusion)*. All *A*-fingers are connected with *Osc1* of Fig. 4.24, and the *B*-fingers are tied to *Osc2*. The *nwell* containing the diode blocks is biased at V_{ctrl}. For simulation, a diode model based on this layout block is used. The model contains a p+/n junction diode and a series resistance. This series resistance Rs_{Diode} is calculated as:

$$Rs_{Diode} = 0.5 * (R_{nwell} * d_{nwell} + R_{Pdiff} * W_F)/(L_F * NF) \qquad (4.30)$$

with:

- R_{nwell} the sheet resistance of the nwell;

- d_{nwell} defined in Fig. 4.25 as the distance between the p+ diode finger and the n+ nwell-strap;

- R_{Pdiff} the resistance of the p+ diffusion;

- W_F the diode finger width;

- L_F the diode finger length;

- NF the total number of parallel diode cells.

The minimal values for d_{nwell} and W_F are dependent on the layout rules of the technology and can be derived automatically within the CYCLONE tool using the *technology-query mode* of the Layout Tool (see Chapter 3).

Like the coil resistance, also the series resistance Rs_{diode} influences the performance of the VCO since it is dissipating energy of the LC tank during the oscillation process and its noise directly contributes to the total noise of the tank. The varactor series resistance is dominated by the distance d_{nwell} between the p+ diode finger and the n+ straps. The total diode resistance is inverse proportional to the *total* diode length. This length is set by the number of parallel connected diode blocks in the final design and thus by the *total* needed diode capacitance. Since only the total diode length is important, the value of L_F in Fig. 4.25 can be chosen freely, e.g. based on the desired aspect ratio for the complete diode cell.

> **Diode cell example:**   As an example, the series resistance and the diode capacitance as a function of the reverse voltage over the diode are derived for a one-finger diode cell in a 0.25μm CMOS technology. In this example $R_{nwell} = 1000\ \Omega/\Box$, $d_{nwell} = 1.8\mu$m, $R_{Pdiff} = 3.7\ \Omega/\Box$, $W_F = 1.2\mu$m and $L_F = 30\mu$m. This results in a diode series resistance of 30.1 Ω for a diode cell with one diode finger. The C/V characteristic for a nominal process and a temperature of 300K is shown in Fig. 4.26. For a tuning voltage that varies between 3 V and 0.6 V, a capacitance ratio of Cmax/Cmin= 1.66 is obtained. The calculated Q of the diode cell is about 55 at 3 GHz.

Also the gain cell has a custom layout, which can be seen in Fig. 4.27. It consists of a bottom part and a top part that are cross connected from drain to gate. Each part has two finger pairs per gain cell block, totaling 4 fingers per part per block. The sources are tied to the ground. The drains from the top part (with *A*-label) are connected to node *Osc1* of Fig. 4.24 using Metal 2, the *B*-labeled drains (bottom part) are connected to *Osc2*. For simulation purposes, the foundry-supplied BSIM3v3 model is used, with addition of a series gate resistance. Also for this building block, the series resistance is important since

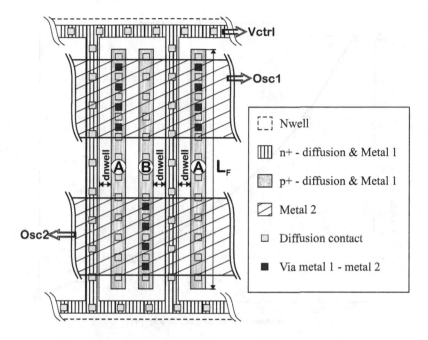

Figure 4.25. Layout of the diode cell

it makes up part of the LC tank. What happens when the gate resistance becomes dominant, is explained in the next subsection.

Gain cell example: As an example, the characteristics of a gain cell with four fingers as in Fig. 4.27 are given for the same $0.25\,\mu$m technology as for the diode cell example above. The finger width W_F is $3.5\,\mu$m, resulting in a total gain cell width of $14\,\mu$m. Obviously, the finger length L is chosen equal to the minimal length of $0.25\,\mu$m. The technology has a polysilicon sheet resistance of $3\ \Omega/\square$. Taking into account the extra polysilicon needed to establish a connection between two fingers and Metal 1, the gate resistance R_{gate} for this block is calculated to be $10.75\ \Omega$. For $V_{gs} - V_t = 0.35$ V, the four finger gain cell draws a current of $780\ \mu$A and has a g_m of 3.36 mS. The total gate capacitance c_{gtot} is 14.7 fF, resulting in an f_T of 36.5 GHz.

As shown in the example above, a single layout block of the gain cell is characterized by choosing $V_{gs} - V_t$, W_F and L, resulting in a certain g_m, bias current and gate resistance R_{gate}, calculated from its physical layout. In fact, the CYCLONE tool automatically derives the g_m and the bias current using a single simulation of the layout block and the gate resistance is derived using the Layout Tool in its *technology-query* mode, as described in Section 3.4.3.

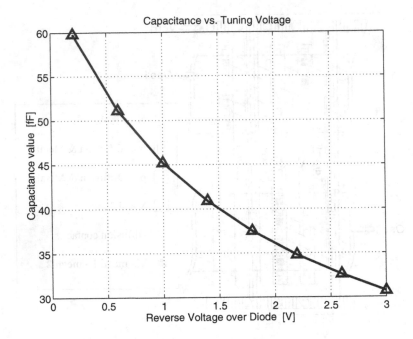

Figure 4.26.   C/V characteristic of the diode cell

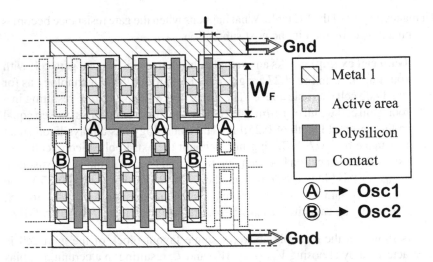

Figure 4.27.   Layout of the gain cell

The total G_m needed to sustain the oscillation is calculated using:

$$G_m = \frac{R_{eff}}{(2\pi f_{osc}L)^2} \tag{4.31}$$

with R_{eff} the effective resistance of the LC tank (as defined in [Cran TCAS95]). From this and the g_m per gain cell block, the number of needed layout blocks N_{gain} is derived. Once the number of gain cells is known, also the total current needed to bias them and the total gate resistance $Rtot_{gate}$ are known.

A final issue concerning the sizing of the layout block of the gain cell is the determination of the finger width W_F. This is discussed next. For the total gate resistance, following equation can be written down:

$$Rtot_{gate} = \frac{R_{gate}}{N_{gain}} \tag{4.32}$$

and it is also clear that following proportionality holds:

$$R_{gate} \propto W_F \tag{4.33}$$

Using

$$N_{gain} = \frac{W_{tot}}{W_F} \tag{4.34}$$

with W_{tot} the total width of the gain cell, (4.32) and (4.33) can be combined into

$$Rtot_{gate} \propto W_F^2 \tag{4.35}$$

Equation (4.35) indicates that the finger width W_F should be kept low in order to minimize the total gate resistance of the gain cell.

The problem with minimizing W_F to a far extent, is that the end-*capacitance* of the gate, caused by the gate extension beyond the active area, is fixed by the topological layout rules. Therefore, the gate capacitance per block consists of a fixed part and a part proportional to W_F. For high-frequency oscillators, the transistor's intrinsic 3dB-frequency should be as close to its f_T as possible. This ensures good gain combined with a low phase shift between gate and drain of the gain cell at the oscillation frequency. For the final design, a finger width of 3.5μm is used. This value is obtained by a manual simulation-based optimization for maximum gain bandwidth of one gain block, loaded with an identical gain block. Since the resistance is proportional to the square of the finger width W_F (4.35), every reduction of W_F leads to a substantial amelioration of total gate resistance.

It seems now that all sizing issues are dealt with. However, still one degree of freedom is left, being the choice of the $V_{gs} - V_t$ value for the gain cell block. This degree of freedom can be used to maximize the symmetry of the waveform [DeMuer ESSCIRC99], but it can also be used to lower the phase noise contribution of the gain cell block. This will be explained at the end of the next subsection.

4.2.3   Power ∝ Phase Noise ?

Knowing that the drain-source current of a MOS transistor is proportional to its g_m, and using (4.31) it follows that the power usage of an oscillator is proportional to the R_{eff} of the LC tank. From the phase noise expression given in (3.9), it follows that also the phase noise is proportional to R_{eff}. The proportionalities formulated above can be further expressed as follows:

$$P \propto R_{eff} \propto Rs_i \tag{4.36}$$

and

$$\mathcal{L}\{\Delta f\} \propto R_{eff} \propto Rs_i \tag{4.37}$$

with Rs_i the series resistance of component i of the LC tank, respectively inductance, diode cell and gain cell.

Combination of the proportionalities above leads to the well-known relationship between power and phase noise for a VCO with an LC tank:

$$P \propto \mathcal{L} \propto R_{eff} \tag{4.38}$$

This means that a lower power comes for free for a lower phase noise. Thus, the best way to design an LC-type oscillator towards low phase noise and low power is to minimize the effective tank resistance.

This design rule is surely valid for LC oscillators for which the inductor is the major source of phase noise within the LC tank. For reasons already indicated above in Section 4.2.1, the resistance in the coil becomes less important than the series resistances of gain cell and varactor diode when the operation frequency of the oscillator is increased to a (very) high value. Let's suppose now a dominance of the gate resistance in the total effective tank resistance. This leads to a power drain that is inverse proportional to the oscillator's phase noise, in contradiction with (4.36). Starting from (4.37) and using (4.32) and (4.33) with a fixed W_F per gain block, the following equation can be derived when the total gate resistance of the gain cell ($Rtot_{gate}$) is the dominant resistance in the LC tank:

$$\mathcal{L}(\Delta f) \propto Rtot_{gate} \propto \frac{1}{N_{gain}} \tag{4.39}$$

Since

$$Itot_{bias} \propto N_{gain} \tag{4.40}$$

with $Itot_{bias}$ the total needed bias current of the gain cell, this means that an increase of power leads to a reduction of the phase noise. Again, this only holds as long as the gate resistance is the prevailing resistance in the LC tank.

As mentioned in the last paragraph of the previous subsection, a value for $V_{gs} - V_t$ still has to chosen. This choice is based on the relative importance of the total gate resistance in the total parasitic resistance of the LC tank. The previous paragraph showed that the total gate resistance should be non-dominant to be able to minimize both phase noise and power by reducing the total series resistance. Reduction of $V_{gs} - V_t$ of the gain cell results in a higher number of gain cells needed for the same value of G_m. This lowers the total gate resistance[2].

Now, all sizing issues of the building blocks of the VCO are dealt with, and a sizing plan can be written. This sizing plan is included as a set of Design Directives within the Design Template of an LC-tank VCO (see Section 3.3.4 for a description of this terminology) in the CYCLONE tool (Section 3.4) that is already mentioned above.

4.2.4 A 17 GHz VCO in 0.25μm CMOS

A 17 GHz VCO design has been made in a standard CMOS 0.25μm technology with 4 Al metal layers and high resistive substrate (15 Ω.cm). The oscillator operates from 16.25 GHz up to 17.715 GHz, when sweeping the control voltage V_{ctrl} from 0.64 V to 1.4 V. This corresponds to a tuning range of 8.6% with a 16.98 GHz center frequency. The tuning characteristic is shown in Fig. 4.28. The VCO has a low phase noise of $\mathcal{L} = -108$ dBc/Hz at an offset of 1 MHz from 17.375 GHz. The phase noise is measured with a PN9000 phase noise measurement set-up, using a downconvertor operating at 15.6 GHz and a phase noise measurement on the downconverted signal using the delay line method. A phase noise plot obtained for an oscillation frequency of 17.375 GHz is shown in Fig. 4.29. All measurements have been done with a power supply voltage of 1.4 V and a bias current for the gain cell of 7.5 mA, resulting in a total power consumption of the VCO core of 10.5 mW. The performance of the VCO is summarized in Table 4.7.

Following the directives given in Section 4.2.1, care has been taken to make the layout of the chip as symmetrical as possible. As explained above, both gain cell and varactor diodes are custom layout to obtain maximum symmetry and minimal parasitics. The high level of symmetry of the chip floorplan can be seen on the chip photograph given in Fig. 4.30. As indicated on this figure, the large metal areas conceal the on-chip decoupling capacitances of V_{dd}, V_{ctrl}, V_{Ibias}, V_{ddbuff} and V_{cmn} of Fig. 4.24. The large decoupling capacitance of V_{cmn} is there to provide an AC-path to ground for the low-frequent 1/f noise of the bias current transistor to prevent 1/f-noise upconversion. Decoupling of V_{ctrl} is needed to ensure a very *clean* voltage at the control node of the varactor diode in order to avoid additional white noise upconversion. To reduce all parasitic resistance in power supply and ground connections, these connections have been made very wide.

---

[2]In this choice of $V_{gs} - V_t$ the influence of 1/f-noise has not been taken into account.

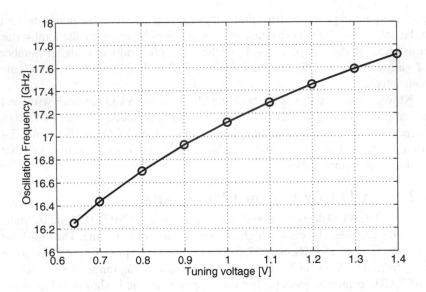

Figure 4.28. Measured tuning range of the VCO

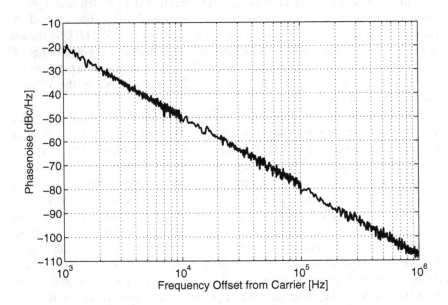

Figure 4.29. Phase noise plot for $F_{osc} = 17.375$GHz

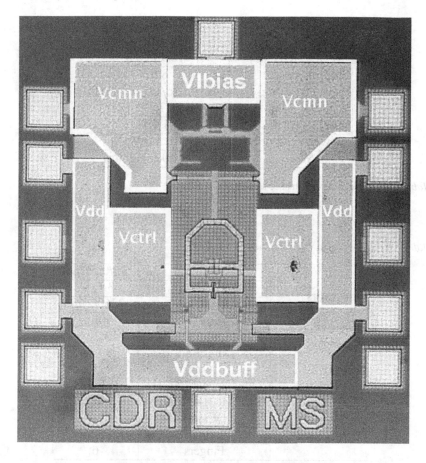

Figure 4.30. Chip photograph of the VCO

Using the CYCLONE tool, a full optimization has been performed on the whole circuit, consisting of inductor, gain transistors and diode. In this optimization, the total effective resistance is minimized, so not only the inductor is optimized. This results in a simulated effective resistance of 5.6 Ω, of which the coil contributes 17%, the diode cell 58% and the gain cell 25%. A good value for $V_{gs} - V_t$ has been derived iteratively running the whole optimization process several times. Finally, a value of 0.35 V has been selected. In Table 4.8 the design values of the transistors, the inductor and the varactor diode are given. For the bias transistors, a longer channel length is chosen to lower the 1/f noise contribution.

Technology	0.25 μm standard CMOS
Substrate resistivity	15 Ω.cm
Center Frequency	16.98 GHz
Tuning Range	8.6%
SSB Phase Noise	-108 dBc/Hz at 1 MHz from 17.375 GHz
Supply Voltage	1.4 V
Power VCO Core	10.5 mW
Chip Size	1.1x1.0(mmxmm)

Table 4.7.   Oscillator performance summary

Table 4.8.   Final design values for the 17 GHz VCO

Transistors	$(W/L)_{gain}$	(70/0.25)μm
	$Wfinger_{gain}$	3.5μm
	$(W/L)_{bias}$	(2360/2)μm
	$Wfinger_{bias}$	40μm
Coil	Width	13.2μm
	Outer Rad.	80μm
	Turns	1
	Inductance@17 GHz	0.32 nH
	Resistance@17 GHz	1.42 Ω
Diode	Width	1.2μm
	Length	30μm
	Fingers	6

4.2.5   Comparison with other Published Work

At the time of its presentation ([DeRa ISSCC01]), the presented chip was the *first* voltage-controlled CMOS oscillator working at frequencies well above 17 GHz (Reference [Klev ISSCC99] in Table 4.9 is a fixed oscillator). Moreover, its specifications can compete with the latest published oscillators, even those fabricated in technologies more suitable for passive components.

4.2.5.1   Table of Comparison

In Table 4.9 the characteristics of a few recently published VCO's are summarized. The phase noise given in the table is recalculated from the reported value to an offset of 1 MHz from the carrier. The power indicated is the power of the gain cell only, with exception of the distributed oscillators, where buffer and core coincide.

Reference	Descr.	F_{osc} [GHz]	Power [mW]	$\mathcal{L}$@1MHz [dBc/Hz]	V_{dd} [V]	Tune [%]	FOM	FOM_T
Craninckx JSSC95	Bondwire	1.8	24	-129	3	5	180	183
Dec ISSCC99	Bondwire	1.9	15	-130	2.5	9	184	190
Svelto CICC00	Bondwire	1.9	2	-120.5	2	15.4	188	196
Soyuer JSSC96	BiCMOS[a]	4	12	-106	3	9	167	172
vdTang JSSC02[b]	BiCMOS	10.0	75	-92	2.7	16	153	161
Jansen ISSCC97	Bipolar[c]	2.2	21.6	-119	2.7	12	173	179
Kinget ESSCIRC98	CMOS/MCM	2.45	5.4	-124	2.7	5	184	187
Kral CICC98	CMOS	1.53	21	-124	3	26[d]	174	184
Liu ISSCC99	coupled	6.53	18	-98	1.5	16.8	162	173
Lam ISSCC99	coupled	5.2	26	-91	2.5	6.8	151	155
Kleveland ISSCC99	CMOS[e]	16.3	52	-110	1.3	0.04[f]	177	*162*
Wu CICC00	CMOS[g]	10.2	35	-114	2.5	12	179	186
Wang ISSCC99	CMOS[h]	9.98	11.6	-115[i]	2.7	3.9	184	186
De Muer CICC00	CMOS	1.82	32.4	-128	1.8	28	182	194
Wang ISSCC01	CMOS	50.1	13	-100	1.3[j]	2.2	183	182
Tiebout ISSCC02	CMOS	51.6	1	-85	1[k]	1.4	179	179
Plouchart ESSCIRC98	SiGe	17.38	22	-105	3.1	3.6	176	177
Plouchart ESSCIRC98	SiGe	6	22	-116	3.1	14.9	178	185
Soyuer JSSC97	SiGe[e]	11	7	-106	3	4.7	178	180
This VCO	CMOS	17.38	10.5	-108	1.4	8.6	183	191

Table 4.9. Recently published integrated oscillators

[a]with thick top metal and thick field oxide
[b]Ring oscillator
[c]high-Q MIS cap and varactor
[d]Switched tuning between multiple inductors
[e]Al/Cu metal layers, thick top metal
[f]Fixed oscillator (distributed amplifier)
[g]Distributed oscillator
[h]Backgate tuning
[i]-118 dBc/Hz has also been reported, but frequency not given; probably for a $V_{ctrl} > V_{dd}$
[j]Maximum value for Vtune is 2.6V
[k]Maximum value for Vtune is 1.6V

For on-chip oscillators, the measured output power off chip or the needed power for the buffers, is irrelevant. In a complete system, the signal of the oscillator is used on-chip, so no power is waisted to buffer the oscillator-signal off chip. The distributed oscillators have the advantage of a large off chip signal power, but it is not clear from the referenced papers [Klev ISSCC99, Wu CICC00] whether or not the power usage can be reduced when only an on-chip oscillator signal is needed. Therefore, the full power for these distributed oscillators is used in the comparison.

4.2.5.2   Figure-of-Merit

To compare these oscillators, two different figures of merit *(FOM)* are used. In the left part of Fig. 4.31 a figure is shown with the FOM as depicted in (4.41) plotted against the reported frequency for the measured phase noise. This is the most frequently used figure of merit for VCO's [King AACD99], with f_0 the center frequency, P the power consumption in mW and $\mathcal{L}$ the phase noise at an offset $\Delta f = 1$ MHz.

$$ \text{FOM} = 10 \log \left(\left(\frac{f_0}{\Delta f} \right)^2 \frac{1}{\mathcal{L}\{\Delta f\}P} \right) \qquad (4.41) $$

To include the tunability of the oscillator, the FOM_T in (4.42) has been introduced in [DeMuer CICC00]. The comparison of recent VCO's based on FOM_T is given in the right part of Fig. 4.31

$$ \text{FOM}_T = 10 \log \left(\left(\frac{f_0}{\Delta f} \right)^2 \frac{1}{\mathcal{L}\{\Delta f\}P} \right) + 10 \log \frac{\text{Tu}}{\text{V}_{dd}} \qquad (4.42) $$

This figure-of-merit includes the VCO-gain, being the tuning-range in percentage (Tu) divided by the power supply voltage (V_{dd}) in volt, normalized to 1 V. A low supply voltage combined with a high tuning range, requires a high VCO gain, rendering the VCO more sensitive to input-noise and making it more difficult to achieve low phase noise. In fact, using FOM_T, the input-referred phase noise of the oscillators is compared.

4.2.5.3   Discussion

In both graphs of Fig. 4.31, the design presented here has a FOM that is ranked rather high compared to other VCO's recently published. Even a comparison with bondwire VCO's and some exotic types is not unfavorable for this VCO.

The comparison with inclusion of tuning range is the most suited, since tuning range is becoming more of a problem with high-frequency oscillators [Wang ISSCC99]. Therefore, Fig. 4.31(b) will be used for comparison. A distinction is made depending on the order of magnitude of oscillation frequency:

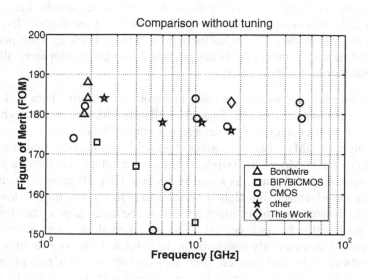

(a) Comparison using (4.41)

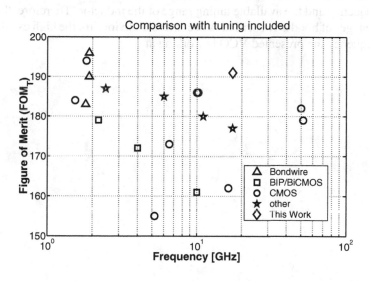

(b) Comparison using (4.42)

Figure 4.31. Comparison of recently published integrated VCO's

- Looking at frequencies within the same order of magnitude, i.e. above 10 GHz, this work outperforms recently published oscillators. Even for frequencies above 2 GHz, it has the highest FOM_T. Looking at the 'normal' FOM, it still is among the better published oscillators with an oscillation frequency higher than 2 GHz.

- Looking at the lower frequencies, this work outperforms both SiGe (silicon germanium) and MCM-based oscillators. Moreover, although they have the lower-frequency advantage, only three VCO's have a comparable or higher FOM_T. Two of those are bondwire oscillators. Comparing this VCO with [DeMuer CICC00] clearly shows the influence of high frequency effects on performance, as stated in Section 4.2.1. Bondwire oscillators have the advantage of a high-Q inductor to reduce phase noise. However, for high-frequency applications a bondwire becomes impractical, because of the small inductance needed. This would result in a very short bondwire with consequently high error on the absolute value, or a lot of parallel bondwires to be implemented. Both implementations would suffer from relatively large parasitics, especially capacitances with possibly low Q. These will deteriorate the overall Q of the inductor and decrease the self-resonance frequency and the available tuning range of the inductor. Therefore, the advantages of bondwire oscillators become less obvious for the high oscillation frequency the presented VCO is working at.

4.3 Conclusions

In this chapter, two totally different types of oscillators have been presented. In a first section, a ringoscillator has been taken under consideration. The use of a four stage ringoscillator has the advantage of a directly available quadrature oscillator signal. It has been demonstrated analytically that the *(4n-1)th* harmonics of a quadrature square wave can be filtered out using a polyphase network. Two chips have been presented that incorporate a ringoscillator linearized using this technique. The first one is a single-channel upconvertor and the second one is a dual-channel upconvertor. Both are tailored for cable applications, with a useful frequency range of 100 MHz to 1.1 GHz. The major disadvantages of a ringoscillator, being a high phase noise and relatively high current consumption have become clear from the measurement results.

In a second section, some design issues concerning oscillators for RF frequencies have been discussed. The need for high symmetry on both floor plan level and building block level has been explained. It has been clarified that in some part of the design space of an LC oscillator the power is inverse proportional to the phase noise. More specifically, this happens when the resistance of the gain cell becomes the dominant source of phase noise in the LC tank. The reported behavior is in this case exactly the opposite of the otherwise generally valid proportionality between phase noise and power of an LC-tank based oscillator. Using the CYCLONE tool presented in Chapter 3, a 17 GHz VCO has been designed. The measurement results have been presented in this chapter and by comparison with other published work, it has been shown that the VCO can be truly considered as state-of-the-art.

Chapter 5

DESIGN OF AN UPCONVERTOR FOR HIGH-SPEED DATA TRANSMISSION

This chapter can be read as the report of a typical design case. Starting from a market projection for a certain type of devices that is quite promising what the expected growth in volume and revenue is concerned, a set of typical system specifications is derived based on the (preliminary) standards that are available. The design then starts with a high-level topology choice, resulting in a block-level diagram. Then the most optimal schematic for each of these different blocks is selected and sized, possibly using CAD tools as described in Chapter 3.

More specifically, the goal is to design an upconvertor capable of transmitting a high-speed data stream on an RF carrier frequency around 2 GHz. "High speed" here means that information is exchanged at a rate that can vary between 384 kbps and 54 Mbps, depending on the technique (or *scheme*) used to modulate the digital data stream and on the speed the wirelessly connected points are moving relative to each other. Typically, this kind of circuits can be used in devices that provide portable wireless data connectivity between e.g. a computer and a wireless access point (WLAN applications) or devices that enable a true mobile data connection between a (slowly) moving handheld device as a PDA or mobile phone and a cellular network acting as an omni-present access point (e.g. UMTS). Following aspects are a common divider of all applications addressed by this design:

1 Since the carrier frequency is situated around 2 GHz, it can be regarded as "medium" high in the RF frequency spectrum used for consumer applications in the year 2002;

2 To transmit the digitally modulated data, an analog signal bandwidth in the order of 5-20 MHz is needed;

3 To ensure a low bit-error rate, the digital modulation scheme requires a
  spectral transmission mask with a Signal-to-Noise (S/N) and Signal-to-
  Distortion (S/D) ratio of more than 30 dBc. This requirement can only
  be met by ensuring a high linearity of the transmitter, which is commonly
  represented by its OIP2 and OIP3.

Finally, it has to be noted that aspects 2 and 3 differentiate the requirements of
this design from those of transmitters used in first and second generation mobile
devices.

   This chapter starts with a short market perspective that indicates the market growth opportunities for integrated circuits that provide high-speed data
communications. A second section clarifies the technical terminology used to
differentiate between broadband/narrow band and high/low-data rate systems
and it also introduces specific system demands for the design. In a third section
the high-level design of the upconvertor and its different building blocks is discussed. A fourth section describes the CMOS implementation of the topology
derived in the third section, including measurements. Then a final conclusion
is formulated.

5.1   Introduction: Market Perspective

   After an exponential growth of the market, the number of GSM subscribers
in Europe has stagnated in the first quarter of 2002 [EMC 02]. Sales numbers
of end-user personal communication devices even showed a negative growth
for 2001 [Cell 02]. As can be seen in Table 5.1, the percentage of replacement
phones is expected to increase to 80% in 2003, and even more later on.

Table 5.1.   Sales Volume of GSM Devices in Europe [Cell 02]

Year	Million	% replacement
99	92	37.2
00	125.9	40.7
01	109.5	62.6
02	110.1	73.4
03	120.7	80.5

   From these numbers, it can be concluded that the market of Personal Communication Systems (PCS) is shifting towards the phase of maturity, with low
growth but high volume sales, and a low profit margin. It is no secret that this
profit margin on devices like GSM's has dropped sometimes even below zero.
   The growth of the PCS consumer market went hand in hand with a steep
progress in chip integration of digital systems and analog front-ends to render
the end-user devices cheap and reliable. The consumer market of high-speed

data communications, of which the high-speed Internet access at home is a very good example, is still in its primary phase with low volumes, low market penetration, high growth expectation and in some cases already a high growth realization (but still low volume). Fig. 5.1 shows a graph with expected volumes of sold devices in the market of high-speed data systems for the next coming years. In Fig. 5.1(a) the predicted number of sold chip sets for Wireless LAN applications is given, while Fig. 5.1(b) depicts the expected number of sold Internet Access Devices. Among these are PC's, mobile phones, Internet set-top boxes and Internet smart appliances[1]. From these graphs it is clear that a serious growth in the number of sold units can be expected in these market segments.

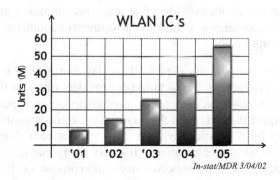

(a) WLAN IC'S

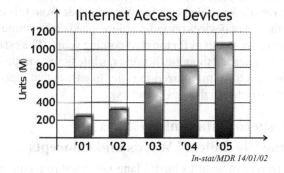

(b) Internet Access Devices

Figure 5.1. Projected market volumes for high-speed data systems

---

[1]E.g. a microwave that downloads a new menu or a soda-vendor machine that reports it's almost empty.

It can be said that the previous boom of hand-held communication devices has been based on a cooperation between two parties, being the technology providers building the devices on the one hand and the voice service providers building the network on the other. To realize a similar exponential growth in the demand of (possibly hand held) devices capable of high-speed data communication, a cooperation between three parties will be required:

- The technology providers that are forced to push the integration level of RF systems even further;

- The data-service providers that are forced to enroll and finance a new network;

- The application-service providers that are expected to design applications for which the consumer will think the investment of another new device and service is worth its price.

This observation can partially explain the delay that is seen in the market-introduction of products by the technology providers (the first party). Moreover, the demands posed on each of the three components or parties involved will be even higher than has been the case for the voice-communication market. Hopefully for all parties involved, this will not be a show-stopper for the next boom in telecommunications. It's very likely that in a few years the market-introduction of the new telecommunication products will be announced by commercials claiming that we succeeded in shrinking even more the distances on our planet...

Now let's leave the planet of marketing and get back to the planet of technology for the remainder of this chapter. We conclude from this overview that the consumer-market of devices providing high-speed data communications has interesting growth perspectives for the next coming years. The system demands are becoming thougher and thougher however, while fierce competition requires the total system cost to be minimized in order to ensure that high revenue will accompany the expected high sales volumes.

5.2   High-speed Data Link Systems

5.2.1   Some "Technical" Words and Concepts ...

Due to an overkill of semi-technical language used in commercials to convince people to buy new devices to get high(er)-speed Internet at home or on their mobile, some confusion sometimes exists when talking about systems incorporating a "high-speed data link", "high bandwidth", "broadband" or "narrow band". Therefore, this chapter starts with a definition of these concepts. The most important issue in the following descriptions is the difference between the *signal bandwidth* and the *frequency range* of the carrier of the signal:

Narrow Band This concept in fact quantifies the *signal* bandwidth of the system it is referring to. Mobile communications systems as GSM and PCS use a signal bandwidth of about 200 kHz to send and receive data between the mobile device and the base station. This is called a Narrow or Small Band(width);

High Bandwidth This is the opposite of Narrow Band. Systems as HFC (cable network) or IEEE802.11 (wireless LAN) use a signal bandwidth of several MHz to send and receive data between the end-user and the base station. *This is often called broadband, causing confusion with the next concept.*;

Broadband System Here the frequency range of the *carrier* is quantified. A cable system as HFC uses a frequency range of 100 MHz up to 1 GHz to send and receive signals (see Section 4.1.1). This is in contradiction with most wireless systems, that operate in small frequency bands of about 10-20% around central frequencies as 900 MHz and 1800 MHz for GSM and DCS-1800 up to 5.6 GHz for wireless LAN;

High-Speed Data Link This quantifies the speed with which data is being sent or received over a data link. Typically, some modulation technique is used to increase the efficiency of a channel. For QAM-type modulation techniques, one can roughly say that the higher the wanted bit rate is, the higher the needed bandwidth and/or the needed signal-to-(noise+distortion) ratio of the link. For spread-spectrum techniques, the bandwidth is determined by the spreading or coding gain.

To summarize: a high-speed data link will need a system that can send a signal with a high bandwidth on a selectable carrier frequency while maintaining a high signal-to-noise and high signal-to-distortion ratios. Whether the frequency range of this carrier is broadband or not has nothing to do with the high or low speed capabilities.

> **Concepts as used in this work** In the previous chapter, some problems occuring within broadband systems are touched by describing the design of a linearized oscillator covering a very broad frequency range. Also, the design of a VCO that operates at a frequency in the application domain of fixed wireless links has been presented, focusing on the problems involved when working with (very) high frequencies in CMOS. In this chapter, the design of an upconvertor for high-speed data transmission is discussed. It is envisaged to be used for both mobile and portable wireless applications. Here the focus lies on the transmission of a high bandwidth.

5.2.2 Some "Real" Numbers for System Demands

In the post-GSM era a few new communication standards for mobile and portable devices are waiting for being introduced into the business and consumer markets. The recently adopted UMTS standard, operating in the 2 GHz range is such a candidate to be selected for the implementation of new *mobile* services on devices like PDA's or cell phones. This standard implements data rates for mobile terminals up to 384 kbps over a bandwidth of 5 MHz using CDMA [UMTS].

For *portable* devices as laptops, generally higher data transmission rates are required. Standards as IEEE 802.15.3 are being developed, providing a data rate of 55 Mbps over a bandwidth of 15 MHz using 64 QAM modulation with Trellis coding [Kara Comm02]. This speed of wireless transmission is also envisaged by the Hiperlan/2 [Hiper2] standard that foresees data rates up to 54 Mbps over a channel 16.25 MHz wide. Here OFDM is implemented using 52 subchannels of which 48 are used as data channels. According to [Kara Comm02], a SNR of more than 20 dB is needed in the receiver to operate at a BER of 10^{-5} or better when receiving a 64 QAM+TCM-modulated signal. The transmitter will need an SNR that is even higher to ensure proper system operation. Of course, any in-band distortion components have to be lower than this SNR and for out-of-band distortion more stringent specifications are formulated. The UMTS standard mentioned in the previous paragraph also defines an indoor wireless transmission speed up to 2 Mbps for non-moving (portable) devices, using also a bandwidth of 5 MHz.

As a comparison: the channel bandwidth of a GSM system is 200 kHz, the needed SNR is about 9 dB and the phase noise specification for GSM is -121 dBc/Hz at a distance of 600 kHz from the carrier([Crols PhD97]). For wireless data communications the allowable phase noise of the oscillator is comparable to this specification. Conclusion of this very brief overview is that in the chip-sets that are to provide high data rates, a signal with a high SNR and a bandwidth some order of magnitudes larger than for first-generation mobile terminals, must be dealt with.

5.3 High-level Design Considerations

In this section a short overview is given of the process that precedes the block level design of the upconvertor. From the brief market study and the rough system specifications given above, a generic formulation of the design goal can be derived:[2]

> The upconvertor must be able to transmit baseband signals with a bandwidth up to 20 MHz on an RF carrier frequency around 2 GHz, while having enough linearity to allow for higher-order digital modulation schemes to be used to generate the baseband signal. Low-power operation is an extra design constraint.

---

[2]The generic aspect is typical for a research project, since not just one specific application or standard is targeted.

Based on expert system knowledge, a topology for a high-bandwidth upconvertor with low in-band distortion is derived. After that, the block-level design can begin. That part is discussed in detail in the next section.

5.3.1 Topology Components

A general topology of the analog part of a transmitter as shown in Fig. 5.2 has following basic components:

- Frequency synthesizer. The frequency synthesizer generates the carrier frequency, and its major performance parameters apart from its power consumption are its phase noise, tuning range and quadrature accuracy;

- Upconversion mixer. The purpose of this component is to perform a frequency translation on the input signal from baseband to carrier frequency. The conversion gain is less important for upconversion, leaving its output distortion, noise floor and quadrature accuracy as major performance parameters;

- Output buffer. Its job is to deliver a "sufficient" output power to drive a power amplifier which could be integrated on the same die, on a different die within the same chip package (multi-chip package) or on another die in a separate package.

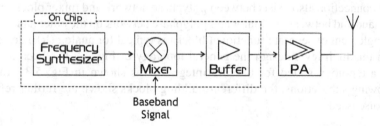

Figure 5.2. General upconvertor topology

At this level, we have to make some decisions that will surely influence the final performance we can obtain with the topology. Since the goal is to fully integrate the system on chip, no external components as filters can be used. As explained in Section 2.4.1 this can be realized using a quadrature upconvertor topology as shown in Fig. 2.20. Thus, a quadrature oscillator and baseband signal has to be provided to the upconversion mixer. The baseband signal is considered to be available as a differential quadrature signal. A differential quadrature oscillator signal has to be provided by the on-chip frequency synthesizer. Since in wireless applications the phase noise has to be low, a ringoscillator solution as discussed in Section 4.1.4 cannot be used here.

In Section 2.4 two commonly used methods to generate a quadrature oscillator signal from a differential input signal have been described:

- A differential VCO operating at twice the wanted frequency, followed by a divider generating a quadrature square-like signal;

- A differential VCO operating at the wanted frequency, followed by a polyphase network to generate a sinusoidal quadrature signal.

Using a direct coupling of the polyphase network between the oscillator and the mixers, avoiding the use of buffers, a very power-optimized topology is obtained. Moreover, the use of a divider would result in the generation of square-like signals. Thus, significant harmonic content is present in the signal coming out of a divider-type quadrature generator. As calculated further on in Section 5.3.5, this harmonic content can result in an unwanted image signal in the output of a linear mixer block. This leads us to the choice for the upconversion mixer block. Because of the high linearity demands, a linear mixer topology is chosen.

A linear mixer has a current output. Therefore, the output buffer has to provide for a low input impedance. As clarified in Section 5.3.4, it also has to incorporate some other functionalities due to the use of a linear mixer. Therefore, the schematic of the output buffer can only be derived after the design of the linear mixer block. In fact, the design of those two blocks is intimately linked to each other. Due to the direct coupling of the polyphase network, a tight connection also exists between polyphase network and mixer block on the one hand and between oscillator and polyphase network on the other hand. This entanglement of separate functional blocks is typical for analog (RF) design.

In this high-level design the general topology of Fig. 5.2 has been refined into a topology suited for on-chip integration as shown in Fig. 5.3. In the following subsections, the different building blocks shown are further refined and discussed.

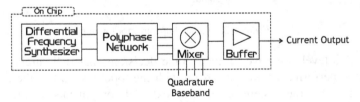

Figure 5.3.   Upconvertor topology suited for full chip integration

5.3.2   Frequency Synthesizer

Generally, a frequency synthesizer consists of two parts: an oscillator with a controllable oscillation frequency and a phase-locked-loop (PLL) to accurately

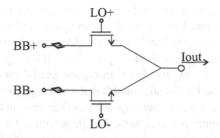

Figure 5.4. Differential linear mixer structure

set the oscillation frequency of the synthesizer to a certain desired value within the frequency range of the synthesizer. For all issues concerning the design of PLLs, the reader is referred to the excellent work of [DeMuer PhD02]. Here, we will focus only on the oscillator within the frequency synthesizer.

For wireless applications, mostly LC-type oscillators are used because of their superior phase noise performance. This is clearly demonstrated in this work, since the phase noise performance of the ringoscillator presented in Section 4.1.4 is much worse than the performance of the LC-tank type oscillators presented in Section 3.4.6 and Section 4.2. Therefore, also for this design an LC-tank type VCO will be used.

5.3.3 Quadrature Generation

For this topic, the reader is referred to the detailed discussion about quadrature generation using polyphase networks in Section 2.4.2.

5.3.4 Linear Mixer block

The basic working principle of a linear mixer is a simple multiplication of the voltages at its drain/source and the voltage at its gate. The behavior of the linear mixer is strongly dependent on the input and output circuitry surrounding the mixer cell. In first order, the output current of a differential linear mixer block as shown in Fig. 5.4 is given by ([Borr JSSC98]):

$$i_{mixer} = \beta \cdot (v_{bb}^2 + 2 \cdot v_{lo} \cdot v_{bb}) \tag{5.1}$$

with v_{bb} the AC component of the baseband voltage and v_{lo} the AC component of the local oscillator signal. The value of β is the constant of the equation describing the I/V behavior of a MOS transistor in the linear region, as given by:

$$i_{DS} = \beta \cdot \left[(v_{GS} - V_T) \cdot v_{DS} - \frac{v_{DS}^2}{2} \right] \tag{5.2}$$

Let's first take a look at the right way to connect the three mixer nodes to the available signals and to the output. In an upconversion mixer, a baseband signal and a high-frequency local oscillator (LO) signal are multiplied to give a high-frequency signal as desired output. Since the output of the mixer is a current, one node is already fixed: the interconnected drain/sources of the two MOS transistors. The baseband signal has a low frequency, and thus efficient buffering of this signal into a low impedance is easier than the buffering of the high-frequent LO-signal into that same low impedance. On the other hand, the gates of the MOS transistors can provide for a high impedance. Thus, in the case of an upconvertor the best thing to do is to use the drain/source node of the mixer transistor as baseband input and the gate node as input for the LO.

To avoid unwanted side-effects, the blocks interfacing with the linear mixer block have to provide for certain conditions. These are explained here. From (5.1) it is clear that any voltage between source and drain of the mixer transistor induces an output current. Therefore, to avoid unwanted components being upconverted, the voltages at the signal input nodes $BB+$ and $BB-$ in Fig. 5.4 should only consist of the clean baseband signal component and the voltage at the output node should be zero in AC. These requirements can be fulfilled by providing a low impedance for all frequencies at the output of the mixers and by providing a low impedance for the high-frequency currents at the input nodes of the mixer. Of course, the RF-current at the output of the mixers is the desired output and has to be separated from the baseband current component.

A DC voltage between the BB-inputs and the output results in a direct upconversion of the LO-signal. Although this upconverted LO is canceled due to the differential construction, it makes the suppression of LO feedthrough mismatch-dependent. By setting the DC voltage over the linear mixers to zero, this unwanted component can be avoided altogether.

Summarizing it can be said that to obtain a good performance of a linear mixer block, the following requirements must be fulfilled [Borr JSSC98]:

- *To avoid LO upconversion:* a zero DC voltage drop between source and drain of the mixer transistors;

- *To avoid upconversion of the baseband frequency f_{BB} to $f_{LO} \pm 2.f_{BB}$:* at the drain/source, a low impedance to ground for DC up to $2.f_{BBmax}$, with f_{BBmax} the maximal baseband frequency;

- *To avoid degeneration of the RF output:* a low impedance for RF signals at the mixers' output and at the mixers' input.

5.3.5   Low LO-Harmonics to avoid Spectral Regrowth

We know already that linear mixers will be used in the upconversion mixer block. In the previous subsection, the needed characteristics of the output buffer following the linear mixer block to certify a correct operation of the mixers, have been summed up. A real implementation of the output buffer will never have infinitely low or high impedances, and it will only maintain its full specifications for a certain frequency range. One could think it to be sufficient for a successful design to have an output buffer that meets the specifications within the frequency range of interest. However, also for frequencies out of this range, the characteristics of the output buffer might influence the final performance of the upconvertor chain.

In this subsection the influence of harmonics present in the quadrature oscillator signal on the upconvertor performance is investigated. More specifically, it is shown that these harmonics can result in spectral regrowth of the wanted signal due to the interaction of the harmonics with the non-ideal behavior of the output buffer following the linear mixer block.

A non-ideal output buffer using some internal feedback to reduce the input impedance will have an input impedance that has a frequency behavior resembling the one shown in Fig. 5.5. In this plot, the application frequency f_c is supposed to be around 2 GHz. Thus, the impedance is low from DC up to frequencies within the application range and then the impedance rises due to the limited gain-bandwidth product of the internal feedback loop. Going towards very high frequencies, the impedance eventually drops again due to parasitics. Although the exact behavior of the input impedance as function of frequency depends on the specific buffer topology that is used, a qualitative discussion based upon Fig. 5.5 is still possible.

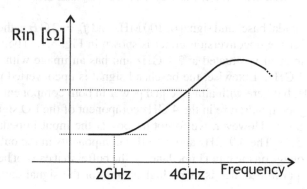

Figure 5.5. Input impedance of a non-ideal buffer

A non-ideal LO signal has some harmonic content. Equation (5.1) is equally valid for these higher harmonics as for the first harmonic, but it has to be noted that parasitic capacitances act as a natural filter for very high frequencies. A differential VCO does not have any even order harmonics. However, the input and output buffers of the polyphase network can introduce second-order harmonics, for instance due to mismatch. Since the second-order harmonic of a 2 GHz LO signal is still well below the f_T of a deep submicron CMOS technology, it is a component that can not simply be neglected. Conclusion is that the baseband signal will also be upconverted to a frequency of $2 \cdot f_c$ and it will have a non-negligible power.

In Fig. 5.6 a plot is shown of the image ratio that can be obtained when the quadrature signal coming out of the polyphase network is used in an ideal quadrature mixer. The image ratio has been tailored to -40 dBc for an application frequency range around 2 GHz. The second harmonic situated at 4 GHz clearly has a theoretical image ratio that is much worse than the in-band image ratio. This is in first order due to the large gain error for frequencies out of the quadrature-generating band of a polyphase network, as can be seen in Fig. 2.30 of Section 2.4.2.8.

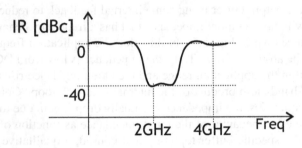

Figure 5.6. Theoretical image ratio after polyphase network

For a sinusoidal baseband signal of 100 MHz and $f_c = 2\,GHz$, the frequency spectrum after the upconversion mixer is shown in Fig. 5.7. The wanted frequency component is situated at 2.1 GHz and has an image with 40 dB less power at 1.9 GHz. Likewise, the baseband signal is upconverted to 3.9 GHz and 4.1 GHz, but here with almost equal power in both components due to the absence of good quadrature in the 4 GHz component of the LO signal. So far, no harm is done. However, we forgot to look at the input impedance of the buffer (Fig. 5.5). The 3.9 GHz and 4.1 GHz components in the output current generate a voltage on the input impedance of the buffer that can not be neglected, since this input impedance is much higher than for the signal components in the 2 GHz range.

This voltage signal at 3.9 GHz is mixed down by the first harmonic of the LO to a frequency of 1.9 GHz and the voltage signal at 4.1 GHz is mixed down by the LO to 2.1 GHz. To understand the deteriorating effect on the upconvertor performance, we have to look at an upconvertor under normal operation. Then, it is not a baseband sine wave that is being transmitted. For a quadrature upconvertor, the baseband signal to be upconverted has an in-phase component

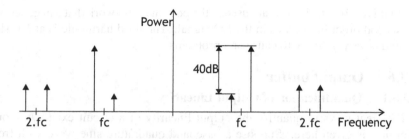

Figure 5.7. Theoretical frequency spectrum after upconversion

I(t) and a quadrature-phase component Q(t). When no quadrature errors are present in the LO signal, the resulting upconverted signal, centered at $\pm f_c$, is given by $I(t) \pm j \cdot Q(t)$ (see Section 2.4.1, equation (2.19)).

Since at $2 \cdot f_c$ the quadrature relation of the LO is really poor, the upconverted baseband signal at the second harmonic of the LO will almost be equal to:

$$I(t) + Q(t) \tag{5.3}$$

This holds for both positive and negative LO frequencies. Thus, it is not possible to recover the I and Q component from this upconverted signal at $2 \cdot f_c$ and it can be treated as a completely distorted signal.

From the explanation above it follows that the presence of higher harmonics in the LO signal can result in a downconversion of this distorted signal to f_c. Thus, an unwanted signal component can be present in the output current of a linear mixer at the LO frequency f_c, resulting in *spectral regrowth* of the wanted output spectrum (see Section 2.4.1). Since the originating distortion signal at $2 \cdot f_c$ is a *common mode* signal for the differential mixer structure of Fig. 5.4, partial cancellation will occur. However, this cancellation is based on a good matching between the two mixer transistors at RF frequencies. Relying on matching, especially the undocumented high-frequency matching between transistors, is a design practice that should be avoided in this kind of building blocks. In this particular case, the best solution is to avoid higher harmonics in the LO signal by careful design of the frequency synthesizer.

Three things are still to be noted:

- The two-channel transmitter presented in Section 4.1.4.2 does not suffer from this problem since the LO is linearized before it is applied to the mixer block;

- The distortion signal at $2 \cdot f_c$ will also be downconverted by the second harmonic of the LO at $2 \cdot f_c$. Since the power of these signals is both small and also here partial cancellation will occur on the resulting baseband current, this effect can be neglected in most cases;

- In this design, no buffers are used in the polyphase network that can generate second-order harmonics in the LO signal. The third harmonic is at 6 GHz and is very unlikely to cause any problems.

5.3.6   Output buffer

5.3.6.1   Quantification of Output Linearity

Different ways to quantify the output linearity of a circuit exist. A short overview is given here. Suppose a baseband quadrature sine wave with frequency f_{BB} is upconverted to an RF frequency f_{LO} using a generic upconvertor as depicted in Fig. 5.3. The wanted output signal then consists of a single-side band sine wave, e.g. at $f_{LO} + f_{BB}$. The output power of this wanted signal component or fundamental tone, is indicated as P_0. This fundamental tone can have distortion components due to non-idealities in the upconversion process. The distortion component at $f_{LO} \pm i.f_{BB}$ is called the i-th harmonic distortion Hi. Mostly, only the second and third-order harmonic distortion components are considered since normally higher harmonics are a lot lower. Expressing the harmonic distortion relative to the power in the fundamental tone leads to following definitions ([Jans PhD01]):

$$\text{HD2} \triangleq \frac{\text{H2}}{\text{P}_0} \quad \text{and} \quad \text{HD3} \triangleq \frac{\text{H3}}{\text{P}_0} \tag{5.4}$$

In real applications, an upconvertor is not used to transmit a single sine wave, but a complete baseband spectrum. To quantify the effect of distortion on the transmitted spectrum, the intermodulation products are used. The result of intermodulation is *spectral regrowth*. If this regrowth is too high, the final spectrum will not fall within the spectral transmission mask that is normally defined by some standard for the considered application.

The intermodulation product is measured by performing a two-tone test. Instead of one single sine wave, now two baseband sine waves with a small frequency difference are applied as input signal. Let the two tones have a frequency of $f_{BB1} < f_{BB2}$. Apart from the normal harmonic distortion components of each tone, also intermodulation products can be found in the upconverted signal. Ordered from close by to further away from the carrier frequency f_{LO}, intermodulation products can be found at following frequencies:

- IMP2: at $f_{LO} \pm (f_{BB2} - f_{BB1})$

- IMP3: at $f_{LO} \pm (2.f_{BB1} - f_{BB2})$

- IMP3: at $f_{LO} \pm (2.f_{BB2} - f_{BB1})$

- IMP2: at $f_{LO} \pm (f_{BB1} + f_{BB2})$

- IMP3: at $f_{LO} \pm (2.f_{BB1} + f_{BB2})$

- IMP3: at $f_{LO} \pm (f_{BB1} + 2.f_{BB2})$

The measured power of the intermodulation products is mostly expressed relative to the power at the fundamental tone:

$$IM2 \triangleq \frac{IMP2}{P_0} \quad \text{and} \quad IM3 \triangleq \frac{IMP3}{P_0} \tag{5.5}$$

Herein, IM2 is called the second-order intermodulation and IM3 the third-order intermodulation.

A common way to express output linearity is the use of the second and third-order output intermodulation intercept points, abbreviated as OIP2 and OIP3. OIP2(3) is defined as the extrapolated output power level P_0 at which the second resp. third-order intermodulation would have the same power as the fundamental tone. Since IM2 and IM3 are expressed relative to P_0, this boils down to the power level that would result in an IM2 resp. IM3 of 0 dB. The extrapolation is done based on intermodulation measurements at power levels that lie within the normal operation range of the output buffer.

Under conditions of low distortion (low power levels), following relationships hold [Sans TCASII99]:

$$IM2 = 2 \cdot HD2 \quad \text{and} \quad IM3 = 3 \cdot HD3 \tag{5.6}$$

$$IM2 \propto P_0 \quad \text{and} \quad IM3 \propto P_0{}^2 \tag{5.7}$$

Eqn. (5.6) relates the harmonic distortion to the intermodulation, and can be used to calculate an estimate of the intermodulation from single-tone measurements. When the power levels are written in dB, following rule-of-thumb can be deduced:

$$IM2 = HD2 + 3.0 \, dB \quad \text{and} \quad IM3 = HD3 + 4.8 \, dB \tag{5.8}$$

Using (5.7), the values of OIP2 and OIP3 can be derived from a single intermodulation measurement. Again expressing the power levels in dB, following well known rules-of-thumb are derived:

$$OIP2 = -(IM2 - P_0) \quad \text{and} \quad OIP3 = -\left(\frac{IM3}{2} - P_0\right) \tag{5.9}$$

These expressions are used to describe the output linearity of the upconvertor that is presented in the next section.

5.3.6.2 Buffer Requirements

To ensure a good performance of the linear mixer block, the output buffer needs a low input impedance for high frequencies. The output of the mixers is a signal current, that must be amplified by the buffer. The output of the buffer must deliver power in an off-chip 50 Ω load. To avoid signal loss in current-to-voltage conversion, a good solution is to use a current amplifier technique for power amplification. An output impedance match can be used to improve the efficiency of the amplifier within a certain frequency range.

5.4 CMOS Implementation of a High-Bandwidth Upconvertor

5.4.1 Introduction

A lot of effort has been spent in the design in CMOS of complete transceiver systems for low data rate, i.e. small band, mobile systems. This resulted in some successful designs in both research and commercial environments [Rofoug JSSC98, Aero, Ajji ISSCC01, Stey JSSC00]. Recent research showed the possible use of a CMOS upconvertor that combines low power consumption with sufficient power output at frequencies up to 2.4 GHz in a system with data rates already an order of magnitude higher than those of first and second generation wireless devices [Zolf ISSCC01].

In this section, an upconvertor is presented that has been implemented in a $0.25\mu m$ CMOS technology ([DeRa ISLPED02]). It is based on the high-level topology that is derived in the previous section and uses some specific circuit techniques within the building blocks that enable the transmission of a baseband signal with a large bandwidth and high SNR, without creating a large increase in power consumption.

In a first subsection, a design overview deals with the global topology of the upconvertor, the used design methodology and the RF-MOST model used for circuit simulations. Using the terminology introduced in the CAD-related chapter Chapter 3, the block-level design methodology would be part of the Design Directives of the Design Template describing the functional block. In the following subsections, the building blocks are reported in detail, starting with the LC-oscillator. The next subsection handles the polyphase network. Section 5.4.5 and Section 5.4.6 deal with the implementation of the linear mixer block and of the output buffer, respectively. Section 5.4.7 discusses the measurements. After that, the conclusions are formulated.

5.4.2 Design Overview

5.4.2.1 Global Topology

Fig. 5.8 shows the block schematic of the implemented upconvertor topology. It is a further refinement on block level of the topology depicted in Fig. 5.3

The first block is an LC-type VCO, that delivers a differential signal to the polyphase network. By this network it is converted into a quadrature LO signal. The input of the polyphase network is coupled capacitively to the oscillator and the output is directly connected to the quadrature mixer block. The baseband

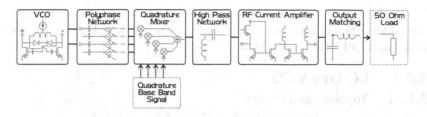

Figure 5.8.   Global upconvertor topology

input of the mixer block is a differential quadrature signal, applied externally. A highpass network acts as an RF/LF splitter at the output of the mixer block to short all low frequency components to ground while passing all RF signals to the RF current amplifier. A 50 Ω output match is implemented as a series inductor and parallel capacitor using on-chip components.

5.4.2.2   Design Methodology

The design methodology used is aimed at a minimization of the total power consumption and consists of a global design on block-specification level and a local optimization on circuit level for each functional block or set of functional blocks. Together with the indication of the specific circuit techniques used, this local optimization will be discussed for each component separately.

5.4.2.3   RF-MOST Model

At the start of the design, no RF-models for the MOS-transistors were available. Therefore, a transistor subcircuit has been constructed using the approach depicted in Section 2.3.2, of which the main aspects are briefly repeated here [3]:

- From the available BSIM3-models an *intrinsic* MOST model is derived by nulling all junction-areas;

- The junctions are added as diodes in the RF-MOST subcircuit, resulting in higher accuracy of the junction capacitor values for low finger numbers;

- Source, drain and substrate resistors are added to the subcircuit;

- A layout-related coupling capacitor between source and drain is added;

- A physical gate resistance is added.

---

[3]In fact, the technology used in the examples of Section 2.3.2 is the same 0.25μm technology as used for the design of the upconvertor

In simulations, the non-quasi-static effect has been found to be negligible for the building blocks and frequencies used in this design. Therefore, it is not added to the RF-MOST subcircuit.

5.4.3   LC-type VCO

5.4.3.1   Topology and Design

The implemented topology for the VCO is shown in Fig. 5.9.

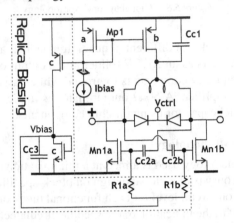

Figure 5.9.   VCO topology with capacitively coupled gain cell transistors

The LC-tank of the oscillator is constructed using an on-chip symmetrical coil and PN-junction diodes of which the capacitance value is controlled by voltage V_{Ctrl}. The inductor has a central tap to which the pMOST bias current source $Mp1b$ is connected. To prevent the 1/f-noise of this transistor to enter the circuit, the central tap is decoupled by the large on-chip capacitor $Cc1$.

The gain cell consists of nMOST pair $Mn1$, of which the gates are coupled to the LC-tank by capacitors $Cc2$. These capacitors effectively decouple the DC output level of the oscillator from the DC biasing voltage of the gain cell. An additional advantage of this topology is that the effective capacitive load of the gain cell on the LC-tank is reduced by a certain factor which is defined by the capacitive divider ratio of the couple capacitance and the parasitic capacitances of the gain cell transistors. Obviously, this capacitive divider simultaneously reduces the effective G_m of the gain cell. Compared to a "standard" topology with directly coupled gain cell transistors as in Fig. 4.24, this topology could be used to:

1 Decrease the current consumption for a certain effective G_m, albeit at the cost of a lower tuning range. This can be established by increasing the width of the gain cell transistors while keeping the coupling capacitance constant. To maintain a certain value for the effective G_m, the current can

be decreased. However, the load of the parasitic capacitance of the gain cell on the LC-tank increases, which reduces the tuning range. A positive effect of the increased transistor sizes is a lower 1/f-noise generation of the gain cell transistors. A negative effect of the lower current is a slightly increased phase noise due to a lower oscillation amplitude;

2 Increase the tuning range of the oscillator, albeit at the cost of a higher current consumption. Increasing the bias current results in a higher g_m of the transistors. The effective G_m of the gain cell can then be reduced by decreasing the coupling capacitance. This results in a reduction of the capacitive load of the gain cell on the LC-tank, leading to a higher tuning range;

3 Decrease the 1/f-noise contribution of the gain cell transistors for a certain current consumption and tuning range. To obtain this, the width of the gain cell transistors is increased. In order to decrease the loading of the LC-tank by the increased parasitic capacitance of the gain cell, the coupling capacitance has to be decreased. This reduces the effective G_m of the gain cell. Since the g_m of the transistors is increased by increasing their width and keeping the bias current constant, the resulting effective G_m can be reduced until it reaches the appropriate value. The increased transistor sizes result in a lower 1/f noise generation. Moreover, when the transistors can be pushed into the weak inversion region by an extreme increase in W/L, this leads to a higher power efficieny and thus a lower current consumption.

In this design, the primary concern was to increase the tuning range of the oscillator, being the second design option in the list above. The power consumption of this design is therefore not expected to be extremely low. As pointed out further on, the VCO has also been designed to have sufficient voltage swing to be coupled capacitively to a polyphase network without the need for extra buffers. Therefore, the total power of quadrature oscillator consisting of VCO and polyphase network will still be "reasonable".

One remaining issue is the gain cell biasing, that requires special care in this topology. Here it is done by a replica biasing of the pMOST current source $Mp1a$ connected in series with nMOST diode $Mn1c$ having the same aspect ratio as the gain cell transistors. The gate of each gain cell transistor $Mn1$ is connected to the resulting replica bias voltage $Vbias$ using a large resistor ($R1$) as depicted in Fig. 5.9. Capacitor $Cc3$ is a low impedance to ground for the noise current of the two resistors. This biasing technique is still apt for further optimization, since it consumes extra current and mismatch can occur between the current flowing through $Mp1a$ and the total current flowing through the gain cell transistors. The current difference caused by this mismatch is forced through the output impedances of transistors $Mn1$ and $Mn1a$, causing the DC output level of the oscillator to be no longer user-controllable if the mismatch becomes too high.

5.4.3.2 Optimization

The optimization of the VCO consists of two parts. First, global optimization of the VCO is done using the in-house CYCLONE tool [DeRa DAC2000] described in Chapter 3. Secondly, an optimal value for resistors $R1$ and capacitor $Cc3$ of Fig. 5.9 is determined, minimizing the influence on the phase noise of the VCO. This optimization has been performed based on circuit simulations using a proprietary phase noise simulator [Desm ESSCIRC97].

Since the DC level at the output of the oscillator can be chosen freely, an optimal value is applied that maximizes the output voltage swing. The result is that no buffers are needed after the VCO to drive the polyphase network. Since the role of these buffers is to provide a low output impedance at RF frequencies, a considerable amount of power can now be saved. Moreover, setting the DC voltage to this optimal value simultaneously maximizes the symmetry of the oscillator signal. According to [Haji JSSC98] this ensures a low upconversion of common-mode 1/f noise (e.g. the 1/f noise current from $Mp1b$).

In Table 5.2 the sizes and values of the components of the final design of Fig. 5.9 are given.

Table 5.2. Sizes and component values for Fig. 5.9

$(W/L)_{Mn1a,b}$	$(140/0.24)\mu m$
$(W/L)_{Mp1}$	$(2200/2)\mu m$
Cc1	40 pF
Cc2	10 pF
R1	100 kΩ
Cc3	40 pF
$Ids_{Mn1a,b}$	3.5 mA
Ids_{Mp1a}	0.7 mA
Ids_{Mn1c}	1.75 mA
Inductor	
Ls @2 GHz	1.7 nH
Rs @2 GHz	0.9 Ω
Radius, Turns, W	$193\mu m$, 3, $44\mu m$

5.4.4 Polyphase Network

A four stage polyphase network is used to generate a quadrature LO signal for a broad range of input frequencies. In one version of the test chip, the polyphase network is driven by an external differential LO to have a larger frequency range for testing. For this block the power consumption is lowered as compared to a traditional topology by omitting the output buffers of the polyphase network and directly coupling the polyphase network to the gates of the mixer transistors.

This direct coupling demands an optimization over three building blocks. The *coupling capacitor* between VCO and polyphase network, the *total network resistance* and the *size of the mixer transistors* all influence at the same time the output current of the mixer block. The goal of the optimization is to maximize this output current, while minimizing the influence of the coupling on the phase noise of the VCO.

The influence of the circuit elements specified above are as follows:

- The polyphase network has a loading effect on the VCO, lowering its oscillation amplitude and deteriorating the phase noise;

- The polyphase network is loaded by the input capacitance of the mixer block, lowering the output amplitude of the polyphase network;.

- The amplitude at the output of the polyphase network directly influences the output current of the mixer block;

- Also the size of the mixer transistors directly influences the output current.

It is clear that the entanglement as described above necessitates a local optimization over a set of functional blocks, being VCO, polyphase network and mixer block.

The values of resistors and capacitors used for the implemented polyphase network are given in Table 2.4 in Section 2.4. A capacitance of 8 pF has been used to couple the polyphase network with the VCO. Fig. 5.10 shows the Image Ratio (IR) of the output signal of the used polyphase network, with and without simulated worst-case mismatch effects for a yield of 99.9%.

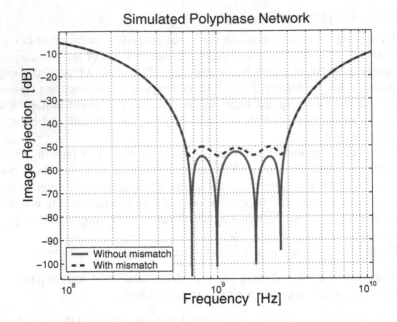

Figure 5.10.   Simulated IR of the generated quadrature signal

5.4.5   Linear Mixer Block

The used mixer topology, including the highpass network at its output, is depicted in Fig. 5.11.

The requirements for a correct operation of the linear mixer block as summed up above in Section 5.3.4 are implemented here as follows:

- The zero DC voltage drop is realized by simply applying a baseband input signal with a zero DC value. (The mixer output is also at DC ground due to the inductor of the highpass network);

- The highpass network realizes a low impedance to ground for low frequencies and a low impedance to the input of the current buffer for high frequencies;

- The input stage of the current buffer realizes a low impedance for RF signals at the mixers' output;

- The on-chip capacitors $Cc1$ in Fig. 5.11 realize a low impedance to ground for RF signals at the mixers' input.

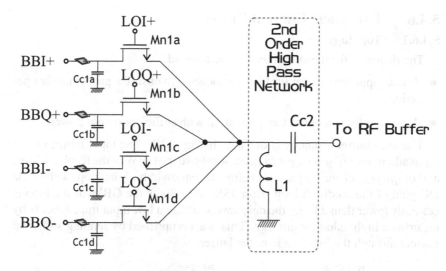

Figure 5.11. Quadrature linear mixer with highpass network

In this building block, optimization is necessary for the on-chip inductor and the coupling capacitor of the highpass network. This optimization is to ensure that the second requirement given above is met and also that low frequencies are restrained from entering the RF current buffer.

A power-efficient implementation is realized by choosing a *passive* network to discriminate between high and low frequencies. An alternative would be to use an active second-order filter. Since the pass-band starts around 900 MHz and goes up to 2 GHz, this solution would consume an amount of power that is comparable to the power usage of the input stage of the RF current buffer. Moreover, the passive solution allows a further enlargement of the transmission bandwidth without power penalty by implementing a higher order passive network, albeit at the cost of a larger area consumption.

In the final design, the mixer transistors have been given the minimal length of 0.24μm and a finger width of 3.5μm. Having 22 fingers, the W/L of the mixer transistor is $(77/0.24)\mu$m. The passive network of Fig. 5.11 has been implemented with a coupling capacitor Cc2 of 20 pF and an inductor L1 of 3.5 nH (at 2 GHz). For the on-chip capacitors Cc1, a value of 20 pF has been chosen.

5.4.6   Two Stage Current Buffer

5.4.6.1   Topology

The demands for the current buffer are twofold:

- A low input impedance for high frequencies to ensure a proper mixer operation;

- A power-efficient current amplification with sufficient output power.

The used buffer topology is depicted in Fig. 5.12. The input buffer has an input admittance of gm_{Mn1} at frequencies above the GBW of the feedback loop, and of $gm_{Mn1}.G$ for frequencies below the bandwidth of the loop, with G the DC gain of the feedback loop [Borr JSSC98]. Since the GBW of the loop is certainly lower than 2 GHz, the only way to obtain a low input impedance is by ensuring a high value for gm_{Mn1}. This is accomplished by sending sufficient current through the input stage of the buffer.

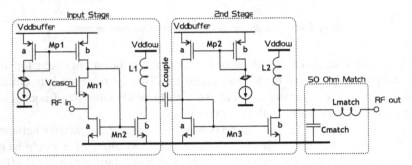

Figure 5.12.   Two stage RF buffer with low impedance input stage

The amplification stages of the current buffer are current mirrors with a ratio $M > 1$. In the first stage, a ratio of 2 is used, and in the second stage a ratio of 3. The first stage is coupled capacitively to the second stage. Thus, pMOST bias current source $Mp2a$ only has to be sized to supply the bias current for $Mn3a$, lowering the parasitic capacitance of $Mp2a$. The (large) mirror current of $Mn2b$ is flowing through an on-chip inductive load connected to Vdd_{low}. The voltage of Vdd_{low} can be made lower than Vdd_{buffer}, because only one transistor has to be set into saturation.

The power usage of this block is minimized by optimization of the load inductance and couple capacitance to obtain a maximum current transfer efficiency of one stage to another.

5.4.6.2   Inductor Optimization

The inductors used as load for the mirror transistors $Mn2b$ and $Mn3b$ in Fig. 5.12 are optimized to obtain maximum current efficiency, in other words, minimum current loss. Therefore, they are designed towards maximum parallel resistance Rp for an inductance value Ls that maximizes the 3 dB bandwidth of the buffer. From circuit simulations, the optimal value for this design is found to be 13 nH. The optimal inductor, having maximum Rp for this value of Ls, is found using a variant of the tool for VCO optimization described in Chapter 3. This tool uses an optimization algorithm (ASA [ASA]) to find the optimal coil geometry by minimization of a cost function. In this case, the cost function CF is constructed as follows:

- $Q_{coil} = 2\pi f_0 Ls/Rs$
- $Rp = Rs(1 + Q_{coil}^2)$
- $\text{Cost}_{Ls} = |Ls_{wanted} - Ls|/Ls_{wanted}$
- $CF = w_1 \cdot \text{Cost}_{Ls} + w_2 \cdot (Rp^{-1})$

Here, f_0 is the maximum frequency of interest, Q_{coil} is the quality factor of the inductor and Ls_{wanted} is the target value for the inductance. Weight factor w_1 is set to a low value if $\text{Cost}_{Ls} < \varepsilon$ with ε the allowable error on the inductance value and to a high value otherwise, while w_2 has a constant medium value. By minimization of this cost function, an inductor geometry is found that has an inductance value "close to" Ls_{wanted} and a maximum value for Rp. In fact, by choosing the weight values and ε appropriately, the optimizer is allowed to make a trade-off between a value for Ls differing slightly from the target and a higher value for Rp.

The optimization targets, parameters and results of the optimization process to find the optimal load inductors for this design are summarized in Table 5.3.

The final circuit of Fig. 5.12 has been sized as depicted in Table 5.4.

5.4.7 Chip Photo & Measurement Results

A photo of the chip, processed in a 0.25μm technology, is given in Fig. 5.13. The used technology provides MMC (metal-metal) capacitors and has a substrate resistivity of 15 Ω.cm. The VCO is situated in the top left corner. Right from it, the four stage polyphase network can be seen. The two coils with small width and five turns are the two load inductors with high Rp from the current buffer. The two other coils are the inductor of the highpass network and of the matching network. The chip measures 1.8x2.3 mm^2

Fig. 5.14 shows the tuning characteristic of the on-chip VCO, as measured on a wire-bonded sample. The oscillator operates for varactor control voltages from 2 V down to 0.15 V, and has an oscillation frequency varying from 1.68 GHz up to 2 GHz, while using 20 mW from a 2 V power supply. The full set of characteristics of the VCO is depicted in Table 5.5.

To demonstrate the linearity of the upconvertor and its usability for high bandwidth applications, measurements have been performed on a flip-chip bonded

Table 5.3. Description of load inductor-optimization process

Optimization Targets	
Frequency	2 GHz
Ls$_{wanted}$	13 nH
Rp	Maximize
Optimization Parameters	
w1 (low/high)	3 / 3000
w2	30000
ε	0.01
Optimization Results	
Ls	13.2 nH @2 GHz
Rs	8.4 Ω @2 GHz
Rp	3.3 kΩ @2 GHz
Q[a]	19.7

[a]unloaded Q, calculated as $Q = \omega Ls/Rs$

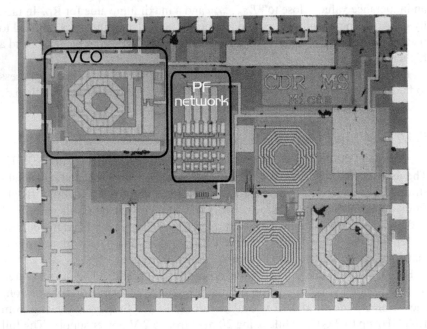

Figure 5.13.   Chip photograph

Table 5.4. Sizes and component values for Fig. 5.12

$(W/L)_{Mn1}$	$(450/0.24)\mu m$
Vcascn	1.87 V
Vddbuffer	2 V
Vddlow	1.3 V
$Vdc_{mixergates}{}^{a}$	0.6 V
$(W/L)_{Mn2a}$	$(90/0.24)\mu m$
$Mirror_{Mn2}$	2
Ids_{Mp1a}	500 μA
$Ids_{Mn1,2a}$	2.5 mA
$(W/L)_{Mn3a}$	$(54/0.24)\mu m$
$Mirror_{Mn3}$	3
Ids_{Mp2b}	280 μA
Ids_{Mn3a}	1.4 mA
Ccouple	20 pF
L1, L2	Ls= 13 nH, Rp= 3.3 kΩ
$(W/L)_{Mp1}$	$(300/0.24)\mu m$
$(W/L)_{Mp2}$	$(180/0.24)\mu m$
Lmatch	3.5 nH
Cmatch	800 fF

[a]see Fig. 5.11

Table 5.5. Measured VCO Specifications

Phase Noise @ 1 MHz for f_{osc}= 2 GHz	-128 dBc
Vdd	2 V
Tuning Range	17%
Center Frequency	1.84 GHz
Total Power Usage	20 mW

sample, driven by an external oscillator. For baseband frequencies f_{BB} up to 33 MHz and an LO frequency f_{LO} of 2 GHz, the harmonic components at $f_{LO} \pm 2.f_{BB}$ and $f_{LO} \pm 3.f_{BB}$ are represented relative to the output power as *HD2* respectively *HD3* in Fig. 5.15. Similar results are obtained at LO frequencies of 900 MHz and 1.5 GHz. Also shown in this plot is the measured

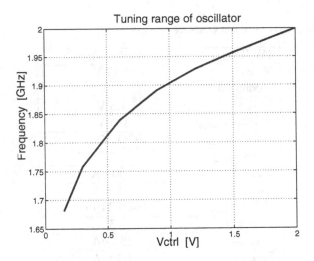

Figure 5.14. Plot of the measured tuning range of the on-chip VCO

output power. For baseband frequencies up to 16.7 MHz, the output power is larger than -12 dBm and up to 33 MHz the distortion is lower than -35 dBc.

As mentioned in Sub-section 5.3.6.1, a more common way to express output linearity is the use of OIP2 and OIP3. These are normally derived from IM2 and IM3 measurements. However, from (5.8) it follows that an estimate of IM2 and IM3 can be calculated directly from single-tone measurements for HD2 and HD3. However, since these higher frequencies suffer from some power loss (see Fig. 5.15), this loss has to be taken into account in the calculation to avoid too optimistic values. Because measurements are only available for baseband frequencies up to 33 MHz, the calculation of IM2 and IM3 can only be performed for baseband frequencies up to 11 MHz. As an example, the method is used here as an approximation for two-tone measurements for a baseband frequency of 5 MHz.

Equation (5.7) shows that one intermodulation (or distortion) measurement theoretically suffices to obtain a value for OIP2 and OIP3. However, for a linear mixer the formula given in (5.7) for IM3 is not valid for "low" values of the output power, according to [Borr JSSC98]. Then, IM3 is simply proportional to P_0. Since measurements of HD2 and HD3 have been performed at one value of P_0 only, no decisive answer can be given to the question whether the measured P_0 should be regarded as "high" or "low". Therefore, the example depicted in Table 5.6 gives both theoretical values for OIP3 for a baseband frequency of 5 MHz. The "normal" one is indicated by $OIP3_3$ and the one valid for "low" power values in a linear mixer is indicated by $OIP3_2$. The value that would be derived from extrapolated measurements will be somewhere in between those values.

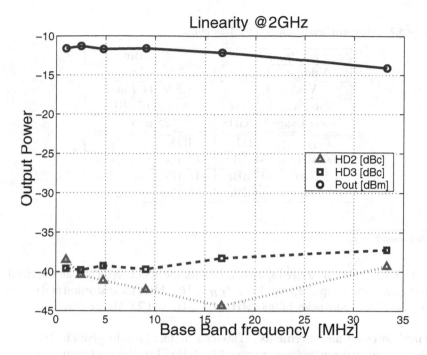

Figure 5.15. Output linearity and power vs. baseband input frequency

Table 5.6. Calculated values for OIP2&OIP3 @2GHz

f_{BB}	5 MHz
OIP2	27 dBm
OIP3$_3$	3.6 dBm
OIP3$_2$	20.2 dBm

From the reasoning above, the representability of OIP3 as single number for the linearity performance of a linear mixer seems questionable. Moreover, an upconvertor will never be used at or in the region of the output power levels indicated by OIP2 or OIP3. However, the intermodulation or distortion at the maximum output power level of the upconvertor really is an important specification. Therefore, the measured harmonic distortion expressed as HD2 and HD3 is used to quantify the performance of this upconvertor. As already mentioned, values for IM2 resp. IM3 can be calculated from these distortion measurements using (5.8).

Table 5.7.　Measured Specifications of the Transmitter

P_{BBin}	-3.5 dBm		
Vdd_{buffer}, I_{buffer}	2 V, 4.7 mA		
Vdd_{low}, I_{low}	1.3 V, 11.7 mA		
Noisefloor @2GHz	< -133 dBc/Hz		
Power Usage @2GHz	25 mW		
$F_{LO}{}^{a}$	HD2	HD3	P_0
900 MHz	-34 dBc	-42 dBc	-10 dBm
1.5 GHz	-37 dBc	-42 dBc	-8.5 dBm
2 GHz	-44 dBc	-38 dBc	-12 dBm

[a]all measured for f_{BB}=16.7 MHz

In Table 5.7 the measured specifications for the upconvertor are summarized, using the same setup as above. For a f_{BB} of 16.7 MHz the distortion and output power at LO frequencies of 900 MHz, 1.5 GHz and 2 GHz are given.

Final notes on measurements　The measurement results given in Table 5.7 show a peak in the output power around 1.5 GHz. The design of the upconvertor has been aimed at an output power of -10 dBm up to a frequency of 2.2 GHz. Clearly, this has not been achieved in the manufactured chip. Several possible reasons can be found to explain this effect:

■ The absence of a measurement-based RF MOST model during simulations and the need to use a construction based model. However, the comparison depicted in Section 2.3.2 did show a good match between a measurement-based model (available after the design had been done) and the used construction-based model;

■ The absence of measurement-based models for the inductors. However, in previous designs the used inductor simulation program (FastHenry) has proven to be quite reliable;

■ The use of underfill underneath the flip-chip bonded die can seriously degrade the performance, as already mentioned in Section 2.5.2;

■ The influence of FIB operations needed to disconnect the on-chip LO and to connect the external LO, on the chip performance has not been investigated.

Due to the fact that many different possible causes exist that could explain the difference between measurements and simulations, it is meaningless to perform a quantitative comparison between possible cause and effect. However, to the author's believe it is the deteriorating influence of the underfill that is to be pointed out as the major cause, since this would be in full accordance with the measured results cited in [Wads ICM98].

5.4.8   Conclusion

In this section, an upconvertor topology has been presented for which measurements show that baseband signals with a bandwidth up to 33 MHz can be upconverted to RF frequencies up to 2 GHz in a standard CMOS technology. A passive highpass network is used to allow the upconversion to be done with a distortion low enough for higher order modulation schemes without the introduction of additional power drain. The power used by the upconvertor is lower than 25 mW. The test chip is shown to keep its functionality at LO frequencies down to 900 MHz.

An on-chip VCO is capacitively coupled to a polyphase network that directly drives the mixer transistors. Thus, power-hungry buffering at RF-frequencies is avoided. The power consumption of the VCO is 20 mW. Specific circuit techniques and local optimization at circuit level of the building blocks have been used during the design to obtain a global power minimization.

The design of the circuit has been done in such a way that the derivation of a set of Design Directives as introduced in Chapter 3 is quite straightforward. A complete redesign of the circuit in a new technology would then take only a fraction of the design time spent on this design. Unfortunately, time to prove this statement lacked in the time-frame of this work. It could be regarded as a first task on the long list of further work-to-be-done....

5.5   Final Conclusion

In this chapter, a typical design case has been described. First the design goal is defined: the design of a transmitter that can be used to send a high-speed data stream on an RF carrier of 2 GHz. From a marketing perspective it becomes clear that this kind of circuits could generate a substantial amount of revenue. Using specifications from different standards in the area of high-speed data communications, a generic set of specifications for a research project is derived. These specifications are used to do a high-level design of the upconvertor. Each of the specified building blocks is then further refined to obtain a circuit schematic. Local optimization of one or multiple building blocks has been used to maximize the performance and minimize the global power consumption. The result is a topology that can be redesigned to specifically target one of the standards or applications used to set up the generic set of specifications. The realization of the test chip shows that the proposed topology integrated in a CMOS technology enables the user to transmit a high-data rate baseband signal on an RF carrier frequency.

Chapter 6

CONCLUSIONS

The design of analog circuits for RF systems has just entered a really exciting time frame. In recent years, silicon-based integrated circuit technologies as CMOS, BiCMOS and SiGe BiCMOS have reached a level of maturity and performance that enables them to be used as core technologies for most of the integrated circuits present in devices for RF applications aiming at consumer markets. By avoiding the use of expensive components manufactured in more exotic technologies as GaAs and by increasing the level of integration which leads to a reduced number of board components, a serious price-cut on the BOM (Bill-of-Materials) of a typical RF system is realized. However, looking at the economic reality at the end of 2002 learns us that it is one thing to have an integrated RF system in-the-pipeline, but it is something completely different to persuade investors into a long-term engagement to support its development until the product becomes profitable. This is caused by the fact that it is very difficult to predict in times of downturn and low sales volumes what kind of chip sets will be the first to see a substantial rise in sales volume when the market starts to recover. Moreover, it is virtually impossible to predict when this recovery will start. From pure technical point of view one can say that the uncertainty about what type of standard will be the next winner in terms of sales and profit, leads to the demand of multi-standard-capable RF front-ends. Therefore, prevailing drivers for further system-integration are this multi-functionality demand and, of course, the total system cost.

An issue that is becoming more and more important within the total cost structure is the choice of the right technology for the integrated circuit within the RF system. One has to balance the reduction of the marginal cost of a single die against the additional NRE cost of a CMOS design including the possible losses due to a delayed market introduction. In the event that this balance is positive for CMOS, still a choice of "flavor" of the technology has to be made. The first chapter clearly showed that a high-resistivity substrate should be chosen since it increases the performance of analog, mixed-signal and high-speed digital circuits. The modeling of MOSFET devices and on-chip passives at RF frequencies is an aspect that must be considered when making the choice of technology provider. Good models often result in fully functional silicon without the need for an expensive redesign. Expensive, because of the price of a new mask set and, again, the cost of a late market introduction. A possible approach to obtain usable models for MOSFET devices and on-chip inductors when no models are provided by the foundry, has been described in Section 2.3.

Accurate on-chip quadrature generation is needed to drive quadrature up-conversion and downconversion mixers that avoid the use of multiple external filters. Two widely-employed methods to obtain a quadrature LO signal on-chip are the use of a polyphase network after a differential VCO and a divider block after a VCO working at twice the wanted LO frequency. Their working principle and possible causes of quadrature inaccuracy have been discussed in Section 2.4.

Some time after a chip has been designed and a layout has been sent to the foundry, the die return. Then, a final manufacturing step has to be performed that secures the die on a substrate or package and realizes a permanent electrical connection between chip and outside world. The connection process itself is called "bonding" and basically two different techniques can be discerned. The wire-bond technique is still used in most cases and consists of a gold or aluminum wire being connected between the bondpads on chip and the conducting leads of the substrate or package. The flip-chip technique uses gold bumps to make the connection between bondpads and substrate leads. In Section 2.5 the importance of inspection after the joining process is demonstrated with X-ray pictures of flip-chip bonded die. This inspection technique has been shown to influence the electrical behavior of the chip, and should therefore be used with caution in a production environment. The increasing system demands of electronic circuits have a direct impact on the daily work of analog designers, especially those doing RF design. Due to the need for system-level design verification while retaining device-level accuracy and the ever increasing time pressure, the daily design effort of an analog designer is fully simulation-based. Although calculations using analytical models are probably better what the gain of insight is concerned, they cannot provide for high accuracy and are mostly seen as merely a waste of time. What such designers might benefit from are design tools that are directly implemented within the daily design environment.

In Chapter 3, an effort is done to define a methodology and a framework that can provide the analog designer with the right environment to put the ever increasing amount of available CPU speed to work for him. The basic goal of the DLE-methodology is to take all repetitive tasks out of the hands of the designer, leaving for him the creative and fun job of topology exploration during his pursuit towards lower power, lower area, higher performance analog circuits. To implement the DLE-methodology, a DLE Tool has been defined that has a Block Modeling Function and a Block Sizing Function. A possible implementation of the latter has been proposed as a Template-Based Design methodology. Wouldn't it be nice that once a promising topology is found, also the Design Directives (or design plan) have been formally captured due to the tools used to find and simulate it. And it get's even better, since the Design Template that consists of the circuit schematic and the Design Directives can be used to automatically optimize the circuit using a Knowledge-Based Performance-Driven optimization process. An example of the proposed implementation has been demonstrated successfully with the CYCLONE tool for the automated design and layout of LC-tank VCO's, described in detail in Section 3.4.

From the market predictions shown in the introduction of Chapter 5, it becomes clear that in the field of high-speed data connections as Internet and wireless LAN a steady growth is expected for the coming years. Therefore, it can be assumed that chip sets tailored to the specific system demands of these applications have a high probability to be among the more profit-bringing products in the near future. For both wireless as wire-line data applications, test chips have been described in this work. In Section 4.1 the problems of broadband systems have been explained and a solution for the upconversion of data in such systems has been proposed under the form of a linear mixer topology combined with a linearized ring oscillator to avoid in-band harmonics. The linearization has been performed with a polyphase network used as a complex filter. For high-frequency applications as fixed-wireless systems and optical systems incorporating high data rates, low-phase noise oscillators are needed. To test the possible integration of such oscillators in CMOS, a 17 GHz VCO has been designed in a standard $0.25\mu m$ CMOS technology. Together with the typical design problems occuring at higher frequencies, it has been presented in Section 4.2.

In the last chapter a lot of the topics presented in the foregoing chapters come together in the design of an RF upconvertor aimed at the transmission of high-bandwidth, high SNR signals. Such signals are typically data signals that have been digitally modulated using some higher-order modulation scheme, hence explaining the higher SNR as compared to first-generation mobile and portable systems. The presented upconvertor operates up to 2 GHz and due to its low power consumption and high-bandwidth capability it might be used in portable and mobile devices of the so-called third or fourth generation.

As final remark it can be said that looking at the results presented in this work, it seems that CMOS is suited for the integration of core building blocks of high data rate devices. This concludes the thesis. A lot of different topics have been touched, some briefly, some more in detail. Hopefully it has been useful. We can end here by uttering our hopes for the future. Let this hope just be that the "seven bad years" that started in 2000 really are nothing but a biblical saying...

6.1 Realized Work

In this work, following chip and design-tool realizations have been described:

- The CYCLONE tool for automated design and layout of LC-type VCO's (Section 3.4);

- A linearized ring oscillator with a high tuning range, made in a 0.5μm CMOS technology (Section 4.1.4);

- A 17 GHz voltage-controlled oscillator made in a 0.25μm technology with high substrate resistivity (Section 4.2.4);

- A 3.3 GHz voltage-controlled oscillator designed using the CYCLONE tool and made in a 0.35μm technology with a low-resistivity substrate (Section 3.4.6);

- A high-bandwidth upconvertor made in a 0.25μm technology with high-resistivity substrate (Section 5.4).

The author also cooperated on the following realizations that are only briefly described or have not been withheld for this text:

- A simulator-optimizer tool for the design of very low phase noise CMOS VCO's [DeMuer ICECS99], being in fact a premature version of the CYCLONE tool;

- The gaRFeeld tool [Vanco DAC00] that uses the same core for inductor simulation as the CYCLONE tool;

- A 1.8 GHz voltage-controlled oscillator designed in a 0.65μm technology with a high-resistivity substrate ([DeMuer ESSCIRC99]);

- A two-channel transmitter for cable applications, designed in a 0.5μm CMOS technology ([Borr ISSCC99, Borr JSSC99], Section 4.1.4);

- Some other voltage-controlled oscillators in the 1.8 GHz range (0.65μm technology) and 10 GHz range (0.25μm technology) that for different reasons were not able to meet the state-of-the art;

- And sometimes it just doesn't work, like that 30 GHz VCO in a fully digital 0.18μm technology...

6.2   Possibilities for Future Research

Very briefly, three topics that should be investigated further are given. They are chosen out of the three most important fields that have been touched in this work: device modeling, design automation and circuit design:

- In the field of device modeling, the TML-coupling between source and drain diffusion areas of a MOSFET device certainly needs some further investigation;

- In the field of design automation, the *Block Modeling Function* of the DLE Tool described in Section 3.2.3 has not been elaborated. In the commercially available Xpedion software package [Xpedion], this modeling functionality is implemented using a cellular neural network. The drawback of this approach is that the training of such a network requires a large data set and since it is a non-linear problem, it can take a long time to derive the optimal coefficients for the individual cells (the so-called *perceptrons*). The use of specific implementations of Support Vector Machines (SVM's) greatly reduces the training effort, due to the fact that it becomes a convex problem that can be solved very efficiently. Since world-leading research is performed in the same building as the research described here has been done[1], it is to be blamed on a lack of time that no effort has been spent on this specific research topic and certainly not on a lack of opportunities;

- In the field of circuit design, the possible topics for further research are widespread. One of the major issues in CMOS upconvertors at frequencies above 2 GHz is the problem of high output power combined with high output linearity. It certainly is a challenge trying to do something in this area between upconvertor and power amplifier.

---

[1]More specifically in the department ESAT-SISTA of the KULeuven

References

[ADA]   Analog   Design   Automation,   *Genius;*
                   http://www.analogsynthesis.com.

[Aero]   Silicon Laboratories, *Aero Si4200*, www.siliconlaboratories.com,
                   2001.

[Ajji ISSCC01]   A. Ajjikuttira, C. Leung, E.-S. Khoo, M. Choke, R. Singh, T.-
                   H. Teo, et al., "A fully-integrated CMOS RFIC for bluetooth
                   applications", in *Digest of Tech. Papers of Int. Solid-State Circuit
                   Conference*, San Francisco, Feb. 2001, pp. 198–200.

[Antrim]   Antrim Design Systems,   *Synthesis & Optimization;*
                   http://www.antrim.com.

[Apar ISSCC02]   V. Aparin, P. Gazzerro, J. Zhou, et al., "A highly-integrated
                   tri-band/quad-mode SiGe BiCMOS RF-to-baseband receiver for
                   wireless CDMA/WCDMA/AMPS applications with GPS capabil-
                   ity", in *Digest of Tech. Papers of Int. Solid-State Circuit Confer-
                   ence*, San Francisco, Feb. 2002, pp. 234–235.

[Arag JSSC99]   X. Aragonès and A. Rubio, "Experimental comparison of substrate
                   noise coupling using different wafer types", *IEEE Journal of Solid-
                   State Circuits*, vol. 34, no. 10, pp. 1405 –1409, Oct. 1999.

[Arch JSSC81]   J. W. Archer, J. Granlund, and R. E. Mauzy, "A broad-band UHF
                   mixer exhibiting high image rejction over a multidecade baseband
                   frequency range", *IEEE Journal of Solid-State Circuits*, vol. SC-
                   16, no. 4, pp. 385–392, Aug. 1981.

[ASA]   L. Ingber,   *Adaptive   Simulated   Annealing   (ASA);*
                   http://www.ingber.com, Caltech Alumni Association, 1993.

[Barcelona]   Barcelona   Design,   *Prado   Synthesis   Platform;*
                   http://www.barcelonadesign.com.

[Beh JSSC01]   F. Behbahani, Y. Kishigami, J. Leete, and A. A. Abidi, "CMOS
                   mixers and polyphase filters for large image rejection", *IEEE*

Journal of Solid-State Circuits, vol. 36, no. 6, pp. 873–887, June 2001.

[Borr ISSCC99]   M. Borremans, C. De Ranter, and M. Steyaert, "A CMOS dual channel, 100 MHz-1.1 GHz transmitter for cable applications", in *Digest of Tech. Papers of Int. Solid-State Circuit Conference*, San Francisco, Feb. 1999, pp. 164–165.

[Borr JSSC98]   M. Borremans and M. Steyaert, "A 2 V, low distortion, 1 GHz CMOS up-conversion mixer", *IEEE Journal of Solid-State Circuits*, vol. 33, no. 3, pp. 359–366, Mar. 1998.

[Borr JSSC99]   M. Borremans, C. De Ranter, and M. Steyaert, "A CMOS dual-channel, 100- MHz to 1.1- GHz transmitter for cable applications", *IEEE Journal of Solid-State Circuits*, vol. 34, no. 12, pp. 1904–1913, Dec. 1999.

[Bout RFD89]   N. Boutin, "Complex signals", *RF-design*, pp. 27–33, Dec. 1989.

[BSIM3]   W. Liu, X. Jin, C. Hu, et al., *BSIM3v3.2.2 MOSFET model, Users' Manual*, University of California, Berkeley, 1999.

[Carc PhD01]   G. Carchon, *Measurement, Modelling and Design of Monolithic and Thin-Film Microwave Integrated Circuits*, PhD Thesis, Katholieke Universiteit Leuven, 2001.

[Carl Spec88]   L. R. Carley and R. A. Rutenbar, "How to automate analog IC designs", *IEEE Spectrum*, vol. 25, pp. 26–30, Aug. 1988.

[Cell 02]   Strategy Analytics, *Handset Sales History & Forecasts*, www.cellular.co.za/news_2002/, 2002.

[Cerny 85]   V. Cerny, "Thermodynamical approach to the traveling salesman problem: An efficient simulation algorithm", *J. Opt. Theory Appl.*, vol. 45, no. 1, pp. 41–51, Jan. 1985.

[Chang PhD95]   H. Chang, *A Top-Down, Constraint-Driven Design Methodology for Analog Integrated Circuits*, PhD thesis, UCBerkeley, 1995.

[Chua 87]   L. O. Chua, C. A. Desoer, and E. S. Kuh, *Linear and nonlinear circuits*, McGraw-Hill, 1987.

[Cohn JSSC91]   J. M. Cohn, D. J. Garrod, R. A. Rutenbar, and L. R. Carley, "KOAN/ANA-GRAMII: New tools for device-level analog placement and routing", *IEEE Journal of Solid-State Circuits*, vol. 26, no. 3, pp. 330–342, Mar. 1991.

[ComCad]   Comcad - Analog Design Support, *Size!; http://www.comcad-ads.de*.

[Cran CICC97]   J. Craninckx and M. Steyaert, "A fully integrated spiral-LC CMOS VCO set with prescaler for GSM and DCS-1800 systems", in *Proceedings Custom Integrated Circuits Conference*, May 1997, pp. 403–406.

[Cran PhD97]   J. Craninckx and M. Steyaert, *Wireless CMOS Frequency Synthesizer Design*, Kluwer Academic Publishers, 1998.

[Cran TCAS95]   J. Craninckx and M. Steyaert, "Low-noise voltage-controlled oscillators using enhanced LC-tanks", *IEEE Transactions on Circuits and Systems II*, vol. 42, no. 12, pp. 794–804, Dec. 1995.

[Crols PhD97]   J. Crols and M. Steyaert, *CMOS Wireless Transceiver Design*, Kluwer Academic Publishers, 1997.

[Crols VLSI96]   J. Crols, P. Kinget, J. Craninckx, and M. Steyaert, "An analytical model for planar inductors on lowly doped silicon substrates for high frequency analog design up to 3 GHz", in *Digest of Technical Papers, Symposium on VLSI Circuits*, Honolulu, June 1996, pp. 28–29.

[Daems TCASI99]   W. Daems, W. Verhaegen, P. Wambacq, G. Gielen, and W. Sansen, "Evaluation of error-control strategies for the linear symbolic analysis of analog integrated circuits", *IEEE Transactions on Circuits and Systems I*, vol. 46, no. 5, pp. 594–606, May 1999.

[Degr JSSC87]   M. Degrauwe, O. Nys, E. Dijkstra, J. Rijmenants, et al., "IDAC: An interactive design tool for analog CMOS circuits", *IEEE Journal of Solid-State Circuits*, vol. SC-22, no. 6, pp. 1106–1116, Dec. 1987.

[DeMuer PhD02]   B. De Muer and M. Steyaert, *CMOS Fractional-N Synthesizers: Design for High Spectral Purity and Monolithic Integration*, Kluwer Academic Publishers, 2003.

[DeMuer CICC00]   B. De Muer, N. Itoh, M. Borremans, and M. Steyaert, "A 1.8 GHz highly-tunable, low-phase-noise CMOS VCO", in *Proceedings Custom Integrated Circuits Conference*, Orlando, May 2000, pp. 585–588.

[DeMuer ESSCIRC00] B. De Muer and M. Steyaert, "A 12 GHz /128 frequency divider in 0.25μmCMOS", in *Proceedings European Solid-State Circuits Conference*, Stockholm, Sept. 2000, pp. 220–223.

[DeMuer ESSCIRC99] B. De Muer, C. De Ranter, and M. Steyaert, "A fully integrated 2 GHz LC-VCO with phase noise of -125 dbc/hz at 600 kHz", in *Proceedings European Solid-State Circuits Conference*, Duisburg, Sept. 1999, pp. 206–209.

[DeMuer ICECS99]   B. De Muer, C. De Ranter, J. Crols, and M. Steyaert, "A simulator-optimizer for the design of very low phase noise CMOS LC-oscillators", in *International Conference on Electronics, Circuits and Systems*, Paphos, Sept. 1999, pp. –.

[DeRa DAC2000]   C. De Ranter, B. De Muer, G. van der Plas, P. Vancorenland, M. Steyaert, G. Gielen, and W. Sansen, "CYCLONE: Automated design and layout of RF LC-oscillators", in *Proceedings IEEE*

Design Automation Conference, Los Angeles, June 2000, pp. 11–14.

[DeRa ECCTD99]   C. De Ranter, M. Borremans, and M. Steyaert, "Design of an integrated transmitter for broadband applications", in *Proceedings European Conference on Circuit Theory and Design*, Stresa, Sept. 1999.

[DeRa ESSCIRC99]   C. De Ranter, M. Borremans, and M. Steyaert, "A wideband linearisation technique for non-linear oscillators using a multi-stage polyphase filter", in *Proceedings European Solid-State Circuits Conference*, Duisburg, Sept. 1999, pp. 214–217.

[DeRa ISLPED02]   C. De Ranter and M. Steyaert, "Design techniques for low power high bandwidth upconversion in cmos", in *Proceedings of International Symposium on Low Power Electronics and Design (ISLPED)*, Monterey, Aug. 2002, pp. 237–242.

[DeRa ISSCC01]   C. De Ranter and M. Steyaert, "A 0.25 μm CMOS 17 GHz VCO", in *Digest of Tech. Papers of Int. Solid-State Circuit Conference*, San Francisco, Feb. 2001, pp. 370–371.

[DeRa TCAD02]   C. De Ranter, G. van der Plas, M. Steyaert, G. Gielen, and W. Sansen, "CYCLONE: Automated design and layout of RF LC-oscillators", *IEEE Transactions on Computer-Aided Design*, vol. 21, no. 10, pp. 1161–1170, Oct. 2002.

[Desm ESSCIRC97]   B. De Smedt and G. Gielen, "Accurate simulation of phase noise in oscillators", in *Proceedings European Solid-State Circuits Conference*, Southampton, UK, Sept. 1997, pp. 208–211.

[Donn PhD98]   S. Donnay, *Analog High-Level Design Automation in Mixed-Signal ASICs*, PhD Thesis, Katholieke Universiteit Leuven, 1998.

[ECD 02]   Eurotraining Course Directory, *TCL Scripting for EDA*, ecd.eurotraining.net, 2002.

[EKV]   M. Bucher, C. Lallement, C. Enz, F. Theodoloz, and F. Krummenacher, *The EPFL-EKV MOSFET Model, Version 2.6; http://legwww.epfl.ch/ekv/*, LEG-EPFL, 1997.

[El-T TCAD89]   F. El-Turky and E. Perry, "BLADES: An artificial intelligence approach to analog circuit design", *IEEE Transactions on Computer-Aided Design*, vol. 8, no. 6, pp. 680–692, June 1989.

[ELEN]   ELEN Division of ESAT, KULeuven, *www.esat.kuleuven.ac.be/elen/research*.

[EMC 02]   EMC World Cellular Database, *Numbers of GSM subscribers (Regional Breakdown)*, www.gsmworld.com, 2002.

[Enz ACD99]   C. Enz and Y. Cheng, *MOS Transistor Modelling Issues for RF Circuit Design* in Analog Circuit Design, Kluwer Academic Publishers, 1999.

[FastHenry] M. Kamon, L. M. Silveira, et al., *FastHenry USER'S GUIDE: version 3.0; ftp://rle-vlsi.mit.edu/pub/fasthenry*, Massachusetts Institute of Technology, 1996.

[Fili ISSCC01] N. Filiol, N. Birkett, J. Cherry, et al., "A 22mW bluetooth transceiver with direct RF modulation and on-chip IF filters", in *Digest of Tech. Papers of Int. Solid-State Circuit Conference*, San Francisco, Feb. 2001, pp. 202–203.

[Fran DATE02] K. Francken, M. Vogels, E. Martens, and G. Gielen, "Daisy-CT: a high-level simulation tool for continous-time $\Delta\Sigma$ modulators", in *Proceedings Design, Automation and Test in Europe Conference and Exhibition*, Paris, Mar. 2002, pp. 1110–1110.

[Free 93] E. M. Freeman, *Magnet 5 User Guide - Using the MagNet Version 5 package from Infolytica*, Infolytica, London and Montreal, 1993.

[Galal TCASII00] S. H. Galal, H. F. Ragaie, and M. S. Tawfik, "RC sequence asymmetric polyphase networks for rf integrated transceivers", *IEEE Transactions on Circuits and Systems II*, vol. 47, no. 1, pp. 18–27, Jan. 2000.

[Geman 84] S. Geman and D. Geman, "Stochastic relaxation, gibbs distribution and the bayesian restoration in images", *IEEE Trans. Patt. Anal. Mac. Int.*, vol. 6, no. 6, pp. 721–741, June 1984.

[Gielen JSSC90] G. Gielen, H. Walscharts, and W. Sansen, "Analog circuit design optimization based on simulation and simulated annealing", *IEEE Journal of Solid-State Circuits*, vol. 25, no. 3, pp. 707–713, June 1990.

[Gielen PhD91] G. Gielen and W. Sansen, *Symbolic analysis for automated design of analog integrated circuits*, Kluwer Academic Publishers, 1991.

[Gielen Proc00] G. Gielen and L. Carley, "Computer-aided design of analog and mixed-signal integrated circuits", *Proceedings of the IEEE*, vol. 88, no. 12, pp. 1825–1854, Dec. 2000.

[Ging EC73] J. Gingell, "Single sideband modulation using sequence asymmetric polyphase networks", *Electrical Communication*, vol. 48, no. 1, pp. 22–25, Jan. 1973.

[Gray 97] P. Gray, W. Hart, L. Painton, C. Philips, M. Trahan, and J. Wagner, *A Survey of Global Optimization Methods*, Sandia National Laboratories, 1997.

[Haji JSSC98] A. Hajimiri and T. H. Lee, "A general theory of phase noise in electrical oscillators", *IEEE Journal of Solid-State Circuits*, vol. 33, no. 2, pp. 179–194, Feb. 1998.

[Harj TCAD89] R. Harjani, R. Rutenbar, and L. Carley, "OASYS: A framework for analog circuit synthesis", *IEEE Transactions on Computer-Aided Design*, vol. 8, no. 12, pp. 1247–1266, Dec. 1989.

[Hers DAC99]   M. del Mar Hershenson, S. S. Mohan, S. P. Boyd, and T. H. Lee, "Optimization of inductor circuits via geometric programming", in *Proceedings IEEE Design Automation Conference*, New Orleans, June 1999, pp. 994–998.

[Hers ICCAD99]   M. del Mar Hershenson, A. Hajimiri, S. S. Mohan, S. P. Boyd, and T. H. Lee, "Design and optimization of LC oscillators", in *Proceedings IEEE/ACM International Conference on Computer Aided Design*, San Jose, Nov. 1999, pp. 550–553.

[Hers TCAD01]   M. del Mar Hershenson, S. P. Boyd, and T. H. Lee, "Optimal design of a CMOS op-amp via geometric programming", *IEEE Transactions on Computer-Aided Design*, vol. 20, no. 1, pp. 1–21, Jan. 2001.

[Hiper2]   European Telecommunications Standards Institute, *HIPERLAN Type 2, System Overview, ETSI TR 101 683*, 2000.

[Hung JSSC02]   C.-M. Hung and K. O, "A fully integrated 1.5-v 5.5-ghz CMOS phase-locked loop", *IEEE Journal of Solid-State Circuits*, vol. 37, no. 4, pp. 521–525, Apr. 2002.

[Ingb CModel92]   L. Ingber and B. Rosen, "Genetic algorithms and very fast simulated reannealing: A comparison", *Mathl. Comput. Modelling*, vol. 16, no. 11, pp. 87–100, Nov. 1992.

[Ingb CModel93]   L. Ingber, "Simulated annealing: Practice versus theory", *Mathl. Comput. Modelling*, vol. 18, no. 11, pp. 29–57, Nov. 1993.

[Jans ACD98]   J. Janssens and M. Steyaert, "Desing of broadband low-noise amplifiers in deep-submicron CMOS technologies", *Analog circuit design. 1 Volt electronics, mixed-mode systems , low-noise and RF power amplifiers for telecommunication*, Kluwer Academic Publishers, Boston, pp. 317–335, 1999.

[Jans EL99]   J. Janssens and M. Steyaert, "Optimum MOS power mathcing by exploiting non-quasi static effect", *IEE Electronics Letters*, vol. 35, no. 8, pp. 672–673, Apr. 1999.

[Jans PhD01]   J. Janssens and M. Steyaert, *CMOS cellular receiver front-ends: From specification to realization*, Kluwer Academic Publishers, 2002.

[Jenei JSSC02]   S. Jenei, B. Nauwelaers, and S. Decoutere, "Physics-based closed-form inductance expression for compact modeling of integrated spiral inductors", *IEEE Journal of Solid-State Circuits*, vol. 37, no. 1, pp. 77–80, Jan. 2002.

[Kamo DAC93]   M. Kamon, M. J. Tsuk, and J. White, "FastHenry: A multipole-accelerated 3-D inductance extraction program", in *Proceedings IEEE Design Automation Conference*, Dallas, June 1993, pp. 678–683.

[Kara Comm02]   J. Karaoguz, "High-rate wireless personal area networks", *IEEE Communications Magazine*, vol. 39, no. 12, pp. 96–102, Dec. 2001.

[King AACD99]   P. Kinget, "Integrated GHz voltage controlled oscillators", in *Proceedings Workshop on Advances in Analog Circuit Design*, Nice, Mar. 1999, pp. 353–381.

[King JSSC97]   P. Kinget and M. Steyaert, "A 1-ghz CMOS up-conversion mixer", *IEEE Journal of Solid-State Circuits*, vol. 32, no. 3, pp. 370–376, Mar. 1997.

[Kirk Science83]   S. Kirkpatrick, C. D. G. Jr., and M. P. Vecchi, "Optimization by simulated annealing", *Science*, vol. 220, no. 4598, pp. 671–680, 1983.

[Klev ISSCC99]   B. Kleveland, C. H. Diaz, D. Vook, L. Madden, T. H. Lee, and S. S. Wong, "Monolithic CMOS distributed amplifier and oscillator", in *Digest of Tech. Papers of Int. Solid-State Circuit Conference*, San Francisco, Feb. 1999, pp. 70–71.

[Knuth 98]   D. E. Knuth, *The Art of Programming, Volume 2: Seminumerical Algorithms*, Addison-Wesley Publishing Company, 1998.

[Koh TCAD90]   H. Y. Koh, C. Séquin, and P. Gray, "OPASYN: A compiler for CMOS operational amplifiers", *IEEE Transactions on Computer-Aided Design*, vol. 9, no. 2, pp. 113–125, Feb. 1990.

[Kol'd SISPAD98]   V. Kol'dyaev, S. Decoutere, R. Kuhn, and L. Deferm, "A comprehensive model of a VLSI spiral inductor derived from the first principle", in *Proceedings International Conference on SImulation of Semiconductor Processes And Devices*, Belgium, Sept. 1998.

[Kouts TCASII00]   Y. K. Koutsoyannopoulos and Y. Papananos, "Systematic analysis and modeling of integrated inductors and transformers in RF IC design", *IEEE Transactions on Circuits and Systems II*, vol. 47, no. 8, pp. 699–713, Aug. 2000.

[Kruis DAC95]   W. Kruiskamp and D. Leenaerts, "DARWIN: CMOS opamp synthesis by means of a genetic algorithm", in *Proceedings IEEE Design Automation Conference*, June 1995, pp. 433–438.

[Lake 94]   K. Laker and W. Sansen, *Design of Analog Integrated Circuits and Systems*, McGraw-Hill, 1994.

[Lamp PhD99]   K. Lampaert, G. Gielen, and W. Sansen, *Analog Layout Generation for Performance and Manufacturability*, KAP, 1996.

[Lee 1990]   E. Lee and D. Messerschmitt, *Digital Communication*, Kluwer Academic Publishers, Boston, 1990.

[Leen 01]   D. Leenaerts, J. van der Tang, and C. Vaucher, *Circuit Design for RF Transceivers*, Kluwer Academic Publishers, 2001.

[Leen ICCAD01]   D. Leenaerts, G. Gielen, and R. A. Rutenbar, "Cad solutions and
                      outstanding challenges for mixed-signal and rf ic design", in
                      *Proceedings IEEE/ACM International Conference on Computer
                      Aided Design*, Oct. 2001, pp. 270–277.

[Levin Spect02]   P. Levin and R. Ludwig, "Crossroads for mixed-signal chips",
                      IEEE Spectrum, vol. 39, no. 3, pp. 38–43, Mar. 2002.

[Long JSSC97]   J. R. Long and M. A. Copeland, "The modeling, characterization,
                      and design of monolithic inductors for silicon RF IC's", *IEEE
                      Journal of Solid-State Circuits*, vol. 32, no. 3, pp. 357–369, Mar.
                      1997.

[Matlab]   The Mathworks, *Matlab*, www.mathworks.com, The Mathworks.

[Meijs Int84]   N. v.d. Meijs and J. T. Fokkema, "VLSI circuit reconstruction from
                      mask topology", *Integration*, vol. 2, no. 2, pp. 85–119, 1984.

[Metr JChem53]   N. Metropolis, A. W. Rosenbluth, A. Teller, and E. Teller, "Equa-
                      tion of state calculations by fast computing machines", *J. Chem.
                      Phys.*, vol. 21, no. 6, pp. 1087–1092, June 1953.

[Model9]   *The   MOS   Model: Level   903;*
                      http://www.semiconductors.philips.com/,   Philips   Semicon-
                      ductors, Eindhoven, 2000.

[Mohan JSSC99]   S. S. Mohan, M. del Mar Hershenson, S. P. Boyd, and T. H. Lee,
                      "Simple accurate expressions for planar spiral inductances", *IEEE
                      Journal of Solid-State Circuits*, vol. 34, no. 10, pp. 1419–1424,
                      Oct. 1999.

[Mori ESSCIRC99]   J. M. et al., "14-bit, 2.2MS/s sigma delta ADCs", in *Proceedings
                      European Solid-State Circuits Conference*, Duisburg, Sept. 1999,
                      pp. 82–85.

[Most TCASII01]   A. H. Mostafa, M. N. El-Gamal, and R. A. Rafla, "A sub-1-V 4-
                      GHz CMOS VCO and a 12.5-GHz oscillator for low-voltage and
                      high-frequency applications", *IEEE Transactions on Circuits and
                      Systems II*, vol. 48, no. 10, pp. 919–926, Oct. 2001.

[NeoLin]   Neolinear, *NeoCircuit(RF); http://www.neolinear.com.*

[Nikn ESSCIRC99]   A. Niknejad and R. Meyer, "Fully-integrated low phase noise
                      bipolar differential VCOs at 2.9 and 4.4 GHz", in *Proceedings
                      European Solid-State Circuits Conference*, Duisburg, Sept. 1999,
                      pp. 198–201.

[Nikn JSSC98]   A. M. Niknejad and R. G. Meyer, "Analysis, design and optimiza-
                      tion of spiral inductors and transformers for Si RF IC's", *IEEE
                      Journal of Solid-State Circuits*, vol. 33, no. 10, pp. 1470–1481,
                      Oct. 1998.

[Nye TCAD88]   W. Nye, D. Riley, et al., "DELIGHT.SPICE: An optimization-based system for the design of integrated circuits", *IEEE Transactions on Computer-Aided Design*, vol. 7, no. 4, pp. 501–519, Apr. 1988.

[Ocho TCAD96]   E. Ochotta, R. Rutenbar, and L. Carley, "Synthesis of high-performance analog circuits in ASTRX/OBLX", *IEEE Transactions on Computer-Aided Design*, vol. 15, no. 3, pp. 273–294, Mar. 1996.

[Ou ACD99]   J. J. Ou, X. Jin, P. R. Gray, and C. Hu, *Recent Developments in BSIM for CMOS RF AC and Noise Modeling* in Analog Circuit Design, Kluwer Academic Publishers, 1999.

[Oust DAC84]   J. Ousterhout, G. Hamachi, R. Mayo, W. Scott, and G. Taylor, "Magic: A VLSI layout system", in *Proceedings IEEE Design Automation Conference*, June 1984, pp. 152–159.

[Papo 84]   A. Papoulis, *Probability, Random Variables and Stochastic Process*, McGraw-Hill, 1984.

[Pawl Comm02]   K. Pawlikowski, H.-D. Jeong, and J.-S. Lee, "On credibility of simulation studies of telecommunication networks", *IEEE Communications Magazine*, vol. 40, no. 1, pp. 132–139, Jan. 2002.

[Pcells]   *Virtuoso Parameterized Cell Reference,version 4.4.3*, Cadence Design Systems, Inc., 1999.

[Pelg JSSC89]   M. J. M. Pelgrom, A. C. J. Duinmaijer, and A. P. G. Welbers, "Matching properties of MOS transistors", *IEEE Journal of Solid-State Circuits*, vol. SC–24, no. 3, pp. 1433–1439, Oct. 1989.

[Pies ISSCC01]   T. Piessens and M. Steyaert, "SOPA: a high-efficiency line driver in 0.35µm CMOS using a self-oscillating power amplifier", in *Digest of Tech. Papers of Int. Solid-State Circuit Conference*, San Francisco, Feb. 2001, pp. 306–307.

[Raza TCASI94]   B. Razavi, R. Yan, and K. Lee, "Impact of distributed gate resistance on the performance of MOS devices", *IEEE Transactions on Circuits and Systems I*, vol. 41, no. 11, pp. 750–754, Nov. 1994.

[Rofoug JSSC98]   A. Rofougaran, G. Chang, J. J. Rael, J. Y.-C. Chang, M. Rofougaran, P. J. Chang, et al., "A single-chip 900 MHz spread-spectrum wireless transceiver in 1 µm CMOS-part I", *IEEE Journal of Solid-State Circuits*, vol. 33, no. 4, pp. 515–534, Apr. 1998.

[Sama ISSCC01]   H. Samavati, H. Rategh, and T. H. Lee, "A fully-integrated 5GHz CMOS wireless-LAN receiver", in *Digest of Tech. Papers of Int. Solid-State Circuit Conference*, San Francisco, Feb. 2001, pp. 208–209.

[Sans TCASII99]   W. Sansen, "Distortion in elementary transistor circuits", *IEEE Transactions on Circuits and Systems II*, vol. 46, no. 3, pp. 315–324, Mar. 1999.

[Silabs]   Silicon Laboratories, *www.siliconlaboratories.com.*

[Spat JSSC02]   A. Spataro, Y. Deval, J.-B. Bégueret, P. Fouillat, and D. Belot, "A
                            VLSI CMOS delay oriented waveform converter for polyphase
                            frequency synthesizer", *IEEE Journal of Solid-State Circuits*, vol.
                            37, no. 3, pp. 336–341, Mar. 2002.

[Stey JSSC00]   M. Steyaert, M. Borremans, J. Jansens, B. D. Muer, N. Itho,
                            J. Craninckx, J. Crols, E. Morifuji, H. S. Momose, and W. Sansen,
                            "A 2 volt CMOS cellular transceiver front-end", *IEEE Journal of
                            Solid-State Circuits*, vol. 35, no. 12, pp. 1895–1907, Dec. 2000.

[Su ISSCC02]   D. Su, M. Zargari, P. Yue, et al., "A 5GHz CMOS transceiver
                            for IEEE 802.11a wireless LAN", in *Digest of Tech. Papers of
                            Int. Solid-State Circuit Conference*, San Francisco, Feb. 200, pp.
                            92–93.

[TCL]   Tcl Developer Xchange, *www.tcl.tk.*

[Tieb ISSCC02]   M. Tiebout, H.-D. Wohlmuth, and W. Simbürger, "A 1 v 51 GHz
                            fully-integrated VCO in 0.12μm CMOS", in *Digest of Tech. Pa-
                            pers of Int. Solid-State Circuit Conference*, San Francisco, Feb.
                            2002, pp. 238–239.

[Tiem ESSDERC99]   L. F. Tiemeijer et al., "MOS model 9 based non-quasi-static small-
                            signal model for RF circuit design", in *Proceedings European
                            Solid-State Device Research Conference*, Duisburg, Sept. 1999,
                            pp. 652–655.

[Torr TCAD96]   A. Torralba, J. Chávez, and L. Franquelo, "FASY: A fuzzy-logic
                            based tool for analog synthesis", *IEEE Transactions on Computer-
                            Aided Design*, vol. 15, no. 7, pp. 705–715, July 1996.

[Tsiv 99]   Y. P. Tsividis, *Operation and Modeling of the MOS Transistor,
                            2nd edition*, McGraw-Hill, 1999.

[UMTS]   European Telecommunications Standards Institute, *Universal Mo-
                            bile Telecommunications System; UE Radio transmission and Re-
                            ception (FDD), ETSI TS 125.101*, 1999.

[Vanco DAC00]   P. Vancorenland and C. De Ranter, "Optimal RF design using
                            smart evolutionary algorithms", in *Proceedings IEEE Design Au-
                            tomation Conference*, Los Angeles, June 2000, pp. 7–10.

[Vanco ICCAD01]   P. Vancorenland, G. van der Plas, M. Steyaert, G. Gielen, and
                            W. Sansen, "A layout-aware synthesis methodology for RF mix-
                            ers", in *Proceedings IEEE/ACM International Conference on
                            Computer Aided Design*, San Jose, Nov. 2001, pp. 358–362.

[vdPlas PhD01]   G. van der Plas, G. Gielen and W. Sansen, *A Computer-Aided
                            Design and Synthesis Environment for Analog Integrated Circuits*,
                            Kluwer Academic Publishers, 2002.

[vdPlas TCAD01]   G. van der Plas, G. Debyser, F. Leyn, K. Lampaert, J. Vanden-
                    bussche, G. Gielen, W. Sansen, P. Veselinovic, and D. Leenaerts,
                    "AMGIE- a synthesis environment for CMOS analog integrated
                    circuits", *IEEE Transactions on Computer-Aided Design*, vol. 20,
                    no. 9, pp. 1037–1058, Sept. 2001.

[Wade 91]   B. C. Wadell, *Transmission Line Design Handbook*, Artech House,
                    1991.

[Wads ICM98]   S. D. Wadsworth, B. J. Buck, A. W. Dearn, A. W. Warner, and
                    I. D. Juland, "Flip-chip GaAs MMICs for microwave MCM-D
                    applications", in *International Conference on Multichip Modules
                    and High Density Packaging*, Denver, Apr. 1998, pp. 273–278.

[Wang ISSCC01]   H. Wang, "A 50 GHz VCO in 0.25 μm CMOS", in *Digest of Tech.
                    Papers of Int. Solid-State Circuit Conference*, San Francisco, Feb.
                    2001, pp. 372–373.

[Wang ISSCC99]   H. Wang, "A 9.8 GHz back-gate tuned VCO in 0.35 μm CMOS",
                    in *Digest of Tech. Papers of Int. Solid-State Circuit Conference*,
                    San Francisco, Feb. 1999, pp. 406–407.

[Wils EE02]   R. Wilson, "World needs fewer chip suppliers, analyst says", *EE-
                    Times*, http://www.eetimes.com/story/OEG20020531S0053, May
                    2002.

[Wu CICC00]   H. Wu and A. Hajimiri, "A 10 GHz CMOS distributed voltage
                    controlled oscillator", in *Proceedings Custom Integrated Circuits
                    Conference*, Orlando, June 2000, pp. 581–584.

[Xpedion]   Xpedion, *www.xpedion.com*, 2001.

[Yue DAC99]   C. P. Yue and S. S. Wong, "Design strategy of on-chip inductors
                    for highly integrated RF systems", in *Proceedings IEEE Design
                    Automation Conference*, New Orleans, June 1999, pp. 982–987.

[Zolf ISSCC01]   A. Zolfaghari, A. Chan, and B. Razavi, "A 2.4 GHz 34 mW
                    CMOS transceiver for frequency-hopping and direct-sequence ap-
                    plications", in *Digest of Tech. Papers of Int. Solid-State Circuit
                    Conference*, San Francisco, Feb. 2001, pp. 418–420.